The Building Safety Act 2022: Questions and Answers

emerald PUBLISHING

ice Publishing

The Building Safety Act 2022: Questions and Answers

A practical approach to design and management of higher-risk buildings

Pat Perry
CMIOSH

Published by Emerald Publishing Limited, Floor 5, Northspring, 21-23 Wellington Street, Leeds LS1 4DL.

ICE Publishing is an imprint of Emerald Publishing Limited

Other ICE Publishing titles:
Fire Safety in Buildings: Questions and Answers
Pat Perry. ISBN: 978-0-72776-647-2
CDM 2015: Questions and Answers: A practical approach to design, safety and wellbeing, Fourth edition
Pat Perry. ISBN: 978-0-72776-584-0
Building Regulations, Codes and Standards: A guide for safe, sustainable and healthy development
Mark Key. ISBN 978-0-72776-535-2

A catalogue record for this book is available from the British Library

ISBN 978-1-83662-467-7

Cover photo: shutterstock_1256193715

Commissioning Editor: Michael Fenton
Content Development Editor: Ryan Molyneux
Books Production Lead: Benn Linfield

Typeset by KnowledgeWorks Global Limited
Index created by David Gaskell
Printed and bound by CPI Group (UK) Ltd, Croydon, CR0 4YY

Contents

Preface

The Building Safety Act 2022 represents a pivotal moment in the evolution of building safety regulations in the UK. This book, structured as a series of questions and answers, aims to provide a thorough understanding of the Act and its implications for various stakeholders in the construction industry.

This book follows the successful format of practical questions and answers found in previous titles such as *Fire Safety in Buildings: Questions and Answers* (2023), *CDM 2015: Questions and Answers: A practical approach to design, safety and wellbeing* (2021), *Health and Safety: Questions and Answers: A practical approach* (2016), and *Risk Assessments: Questions and Answers: A practical approach* (2017).

The Grenfell Tower fire in June 2017 was one of the UK's most devastating modern disasters and serves as a tragic backdrop to the necessity for reform. The fire exposed significant failures in fire safety measures, resulting in 72 fatalities and numerous injuries. The subsequent public inquiry revealed critical shortcomings in building safety regulations, underscoring the urgent need for comprehensive reforms.

The Building Safety Act 2022 is a new regulatory framework designed to address these failures and enhance safety standards, particularly for high-risk residential buildings. Key provisions include the establishment of the Building Safety Regulator, the introduction of Accountable Persons and the implementation of enforcement sanctions such as stop notices and compliance notices. The Act also emphasises the importance of competency standards, and the need for effective communication and collaboration among various stakeholders involved in building safety management.

This book covers various aspects of the Building Safety Act 2022, including the roles and responsibilities of Clients, Principal Designers and Principal Contractors, as well as fire safety measures in high-rise residential buildings. It provides guidelines on managing safety risks, outlines the concept of the Safety Case and the Safety Case Report, and discusses the three gateways designed to enhance safety standards during the design, construction and occupation phases. It also covers the golden thread of information and addresses mandatory occurrence reporting, the new change control regime, residents' engagement strategy and the significance of construction products in relation to building safety.

By offering a pragmatic approach to conveying the information needed to comply with the legislation, this book aims to equip a broad range of professionals – including engineers, architects, project managers, company directors and health and safety professionals – with the knowledge and tools necessary to navigate the complexities of the Building Safety Act 2022 and contribute to the creation of safer built environments.

Pat Perry CMIOSH
April 2025

About the author

Pat Perry CMIOSH qualified as an environmental health officer in 1978 and spent the early years of her career in local government in the UK enforcing environmental health regulations, particularly health and safety law, which became her passion. Pat left local government and established a very successful Environmental Health Consultancy, which was subsequently integrated into a large regulatory compliance group. She has extensive knowledge of the retail, commercial and hospitality sectors and has served on various working parties on both health and safety and food safety issues. Pat regularly contributes to professional journals and appears as a speaker at related conferences and seminars. Her previous publications include *Fire Safety in Buildings: Questions and Answers* (2023), *CDM 2015: Questions and Answers: A practical approach to design, safety and wellbeing* (2021), *Health and Safety: Questions and Answers: A practical approach* (2016) and *Risk Assessments: Questions and Answers: A practical approach* (2017).

The Building Safety Act 2022: Questions and Answers: A practical approach to design and management of higher-risk buildings

Pat Perry
ISBN 978-1-83662-467-7
https://doi.org/10.1108/978-1-83662-464-620251001

Chapter 1
The Grenfell Tower fire

What was the Grenfell Tower fire?

The fire that destroyed Grenfell Tower in June 2017 was one of the UK's worst modern disasters.

Just before 01:00 on 14 June, fire broke out in the kitchen of a fourth floor flat at the 23-storey tower block in North Kensington, West London.

Within minutes, the fire had accelerated up the exterior of the building and then spread to all four sides. By 03:00, most of the upper floors were engulfed in flames.

Seventy-two people died and many suffered severe injuries from burns and smoke inhalation.

A fire in the multi-storey building was not expected to spread as rapidly as it did. Such speed took the emergency services by surprise and they were ill equipped to deal with it.

How did the fire start?

The fire began due to an electrical fault 'in or around' a Hotpoint fridge-freezer in Flat 16 on the fourth floor of the tower block.

The occupier immediately called the fire brigade, which logged the call at 00:54.

Four fire engines were sent to the scene, the first arriving at 00:59.

The first fire crew entered the flat at 01:07 – an approximate timing taken from a thermal imaging camera. They initially searched the bedrooms and did not enter the kitchen until 01:14.

In the kitchen, a firefighter described seeing an 'isolated curtain of flame from about 2–3 feet in the air to the ceiling'.

Thermal images captured by the fire crew appear to show that 'hot fire gases and flames had spread across the window space'.

As the fire crew dealt with the fire in the fridge-freezer, their thermal images also appear to show falling embers outside the kitchen window.

Mobile phone images taken by the tenant from outside Grenfell Tower approximately 11 min after his first 999 call show an orange glow of flames around the kitchen window, and later a fire burning more intensely in the area of the window filler panel and extractor fan.

Subsequent photos suggest that the 'fire was continuing to develop and grow', by 01:09 becoming 'external to the building'.

Although the timings provided are approximate, it appears that the fire had spread to the cladding before the firefighters had entered the kitchen.

How did the fire spread?

From the fourth floor, the fire spread rapidly upwards and across the eastern side of the building. From there, it spread across the north face of the tower.

Mobile phone videos show the blaze reaching the top floor on the east side of Grenfell Tower by about 01:26, less than 30 min after firefighters had arrived.

In a report to the Grenfell Public Inquiry, fire safety engineer Dr Barbara Lane identified the fire spreading vertically up the tower columns, and 'laterally along the cladding above and below the window lines (and) the panels between windows'.

The fire had spread to the north side of the tower by 01:42, Dr Lane recorded.

At 01:52, the fire also began travelling across the eastern side towards the south in the other direction.

At 02:06, London Fire Brigade declared the fire a 'major incident'. At this point, some 40 fire engines were either at or en route to the scene.

Grenfell Tower had a 'stay put' fire policy – essentially, the building design would contain a fire in a single flat for as long as it took fire crews to bring it under control.

On the night, many residents were told to remain in their flats by the emergency services, only to become trapped as the fire blazed out of control and thick, poisonous smoke spread up the single narrow stairwell.

Dr Lane said that the stay put policy had 'substantially failed' by 01:26 – less than 30 min after the first firefighters were at the scene.

Some people ignored the stay put advice and made it down the stairs to safety.

A total of 65 people were rescued from the building by firefighters.

Some residents went upwards and sought refuge in flats of friends and neighbours on the upper floors. Twenty-four people died on the top floor of the tower block.

By 02:10, multiple internal fires could be seen burning inside the building.

At 02:22, fire had spread to the south side of the tower and by 02:30 it was reported that the eastern side of the building was 'fully involved in fire'.

The stay put advice was finally abandoned at 02:47, when the incident commander gave the order to 'advise people to make efforts to leave the building'.

Counsel to the Grenfell Inquiry, Richard Millet QC, told the Public Inquiry hearing that 144 people managed to evacuate before 01:38, but only 36 after the stay put guidance was abandoned.

By 02:51, the fire had reached the western side. At this point, some 63 flats were on fire and more than 100 people remained in the building.

At 04:30, the whole building was engulfed, with more than 100 flats on fire.

The blaze did not burn itself out until 01:14 BST on Thursday – 24 hours later.

What caused the fire to spread so quickly?

The most significant part of the renovation of Grenfell Tower was the addition of external cladding. This consisted of aluminium sheets bonded to a central plastic (polyethylene) core (aluminium composite material (ACM)) cladding.

In his report to the Public Inquiry, Professor Luke Bisby said evidence 'strongly supports' the theory that the polyethylene material in the cladding was the primary cause of the fire's spread.

> The ACM (aluminium composite material) product on Grenfell Tower incorporates a highly combustible polyethylene polymer filler which melts, drips, and flows at elevated temperature. The polyethylene filler material is expected to release large amounts of energy during combustion.

Images after the fire show evidence of melted cladding on the tower's columns. His report also suggests that vertical cavities within the cladding structure played a role in the spread of the fire, as did the insulation, although evidence was inconclusive.

A number of other flammable materials, including a polyurethane polymer foam insulation board that pre-dated the refurbishment were also present and may have contributed, Professor Bisby found.

In her report, fire expert Dr Barbara Lane identifies combustible materials used in the refurbishment of the tower's windows as another factor in allowing the fire to spread.

The Grenfell Inquiry heard that the building's smoke extraction system was not working, and that firefighters experienced problems with the water supply because there were no 'wet risers' – water-filled pipes running up the building to be used in the event of fire.

Giving evidence to the Inquiry, Dr Lane said exposed gas pipes installed in 2016 were another contributory factor and none of the flat doors met current fire resistance standards.

Work done on the lifts in 2005 and 2012–2016 left them unfit for evacuating vulnerable residents and aiding the emergency response, Dr Lane said.

Numerous fire protection measures failed and stair lobbies became full of toxic smoke.

Glass in windows failed due to the heat generated by the fire and allowed flames to penetrate flats.

A catalogue of failures in respect of fire prevention and fire precautions caused the fire to spread rapidly with devastating consequences.

The Building Safety Act 2022: Questions and Answers: A practical approach to design and management of higher-risk buildings

Pat Perry
ISBN 978-1-83662-467-7
https://doi.org/10.1108/978-1-83662-464-620251002

Chapter 2
The Public Inquiry

Why was a Public Inquiry set up in relation to the Grenfell Tower fire?

The Grenfell Tower fire was a major catastrophe causing the loss of many lives. It should not have happened.

The Inquiries Act 2005 provides a comprehensive statutory framework for inquiries set up by Ministers to look into matters of public concern. The Grenfell Tower fire was a matter of public concern. It had caused outrage among the public. People wanted the circumstances of the fire investigated and they wanted answers as to why it happened. People wanted those responsible to be held accountable for their failings.

The Prime Minister at the time announced on the day after the fire that there would be a Public Inquiry.

The Inquiry's Terms of Reference were

(i) to examine the circumstances surrounding the fire at Grenfell Tower on 14 June 2017, including

- (a) the immediate cause or causes of the fire and the means by which it spread to the whole of the building;
- (b) the design and construction of the building and the decisions relating to its modification, refurbishment and management;
- (c) the scope and adequacy of building regulations, fire regulations and other legislation, guidance and industry practice relating to the design, construction, equipping and management of high-rise residential buildings;
- (d) whether such regulations, legislation, guidance and industry practice were complied with in the case of Grenfell Tower and the fire safety measures adopted in relation to it;
- (e) the arrangements made by the local authority or other responsible bodies for receiving and acting upon information either obtained from local residents or available from other sources (including information derived from fires in other buildings) relating to the risk of fire at Grenfell Tower, and the action taken in response to such information;
- (f) the fire prevention and fire safety measures in place at Grenfell Tower on 14 June 2017;

(g) the response of the London Fire Brigade to the fire; and
(h) the response of central and local government in the days immediately following the fire; and

(ii) to report its findings to the Prime Minister as soon as possible and to make recommendations.

Sir Martin Moore-Bick was appointed Chairman of the Inquiry. He was a former Court of Appeal Judge and was deemed by the Lord Chancellor to be suitable for the post. He could start the Inquiry immediately thus forestalling any criticisms of the government delaying its response to the fire.

Sir Moore-Bick set the agenda for the Inquiry and determined its timetable. He appointed a team of officials to assist him and assumed the powers of the Inquiries Act 2005 to demand evidence be submitted from whomever he deemed necessary.

How long did the Public Inquiry take to finalise all hearings and publish its final report?

Sir Martin Moore-Bick acknowledged in a statement that it had taken 7 years for the Inquiry into the disaster to be concluded because of the number of failings it had uncovered and the time it had taken to investigate them all. 'As the investigation progressed, it uncovered many more matters of concern,' he said.

When were the reports published?

The Public Inquiry reported in two stages. The Phase One Report was published in late October 2019 and the final, Phase Two Report was published in late September 2024.

What was the outcome of the Public Inquiry?

Both reports made recommendations to the government about shortcomings in legislation and enforcement procedures, emergency services responses to major incidents, construction industry standards, design and management of high-rise buildings and numerous miscellaneous matters.

Fire safety laws were strengthened, Building Regulations were amended, building safety systems were introduced and a new regulatory framework was introduced.

The Building Safety Act 2022 was passed and came in to force in stages from 2022.

High-rise, high-risk residential buildings were to be controlled for the first time.

What were the recommendations in the Inquiry Phase One Report?

Phase One of the Inquiry Report made 46 recommendations relating to key topics explored during the Inquiry hearing.

Many of them have been actioned by the government in association with the National Fire Chiefs Council and London Fire Brigade.

Recommendation 33.10a

[References reflect the numbering from the Inquiry Report, i.e. Chapter 33.]

That the owner and manager of every high-rise residential building be required by law to provide their local fire and rescue service with information about the design of its external walls together with details of the materials of which they are constructed and to inform the fire and rescue service of any material changes made to them.

Recommendation 33.10b

That all fire and rescue services ensure that their personnel at all levels understand the risk of fire taking hold in the external walls of high-rise buildings and know how to recognise it when it occurs.

Recommendation 33.11a

That the London Fire Brigade review, and revise as appropriate, Appendix 1 to Policy Number 633 to ensure that it fully reflects the principles in Generic Risk Assessment 3.2.

Recommendation 33.11b

That the London Fire Brigade ensure that all officers of the rank of Crew Manager and above are trained in carrying out the requirements of Policy Number 633 relating to the inspection of high-rise buildings.

Recommendation 33.12a

That the owner and manager of every high-rise residential building be required by law to provide their local fire and rescue services with up-to-date plans in both paper and electronic form of every floor of the building identifying the location of key fire safety systems.

Recommendation 33.12b

That the owner and manager of every high-rise residential building be required by law to ensure that the building contains a premises information box, the contents of which must include a copy of the up-to-date floor plans and information about the nature of any lift intended for use by the fire and rescue services.

Recommendation 33.12c

That all fire and rescue services be equipped to receive and store electronic plans and to make them available to incident commanders and control room managers.

Recommendation 33.13a

That the owner and manager of every high-rise residential building be required by law to carry out regular inspections of any lifts that are designed to be used by firefighters in an emergency and to report the results of such inspections to their local fire and rescue service at monthly intervals.

Recommendation 33.13b

That the owner and manager of every high-rise residential building be required by law to carry out regular tests of the mechanism which allows firefighters to take control of the lifts and to inform their local fire and rescue service at monthly intervals that they have done so.

Recommendation 33.14a

That the London Fire Brigade review its policies on communications between the control room and the incident commander.

Recommendation 33.14b

That all officers who may be expected to act as incident commanders (i.e. all those above the rank of Crew Manager) receive training directed to the specific requirements of communication with the control room.

Recommendation 33.14c

That all control room operators of Assistant Operations Manager rank and above receive training directed to the specific requirements of communication with the incident commander.

Recommendation 33.14d

That a dedicated communication link be provided between the senior officer in the control room and the incident commander.

Recommendation 33.15a

That the London Fire Brigade's policies be amended to draw a clearer distinction between callers seeking advice and callers who believe they are trapped and need rescuing.

Recommendation 33.15b

That the London Fire Brigade provide regular and more effective refresher training to control room operators at all levels, including supervisors.

Recommendation 33.15c

That all fire and rescue services develop policies for handling a large number of fire survival guidance calls simultaneously.

Recommendation 33.15d

That electronic systems be developed to record fire survival guidance information in the control room and display it simultaneously at the bridgehead and in any command units.

Recommendation 33.15e

That policies be developed for managing a transition from 'stay put' to 'get out'.

Recommendation 33.15f

That control room staff receive training directed specifically to handling such a change of advice and conveying it effectively to callers.

Recommendation 33.16

That steps be taken to investigate methods by which assisting control rooms can obtain access to the information available to the host control room.

Recommendation 33.17

That the London Ambulance Service and Metropolitan Police Service review their protocols and policies to ensure that their operators can identify fire survival guidance calls (as defined by the London Fire Brigade) and pass them to the London Fire Brigade as soon as possible.

Recommendation 33.18a

That the London Fire Brigade develop policies and training to ensure better control of deployments and the use of resources.

Recommendation 33.18b

That the London Fire Brigade develop policies and training to ensure that better information is obtained from crews returning from deployments and that the information is recorded in a form that enables it to be made available immediately to the incident commander (and thereafter to the command units and the control room).

Recommendation 33.19

That the London Fire Brigade develop a communication system to enable direct communication between the control room and the incident commander and improve the means of communication between the incident commander and the bridgehead.

Recommendation 33.20

That the London Fire Brigade investigate the use of modern communication techniques to provide a direct line of communication between the control room and the bridgehead, allowing information to be transmitted directly between the control room and the bridgehead and providing an integrated system of recording fire survival guidance information and the results of deployments.

Recommendation 33.21a

That the London Fire Brigade urgently take steps to obtain equipment that enables firefighters wearing helmets and breathing apparatus to communicate with the bridgehead effectively, including when operating in high-rise buildings.

Recommendation 33.21b

That urgent steps be taken to ensure that the command support system is fully operative on all command units and that crews are trained in its use.

Recommendation 33.22a

That the government develop national guidelines for carrying out partial or total evacuations of high-rise residential buildings, such guidelines to include the means of protecting fire exit routes and procedures for evacuating persons who are unable to use the stairs in an emergency, or who may require assistance (such as disabled people, older people and young children).

Recommendation 33.22b

That fire and rescue services develop policies for partial and total evacuation of high-rise residential buildings and training to support them.

Recommendation 33.22c

That the owner and manager of every high-rise residential building be required by law to draw up and keep under regular review evacuation plans, copies of which are to be provided in electronic and paper form to their local fire and rescue service and placed in an information box on the premises.

Recommendation 33.22d

That all high-rise residential buildings (both those already in existence and those built in the future) be equipped with facilities for use by the fire and rescue services enabling them to send an evacuation signal to the whole or a selected part of the building by means of sounders or similar devices.

Recommendation 33.22e

That the owner and manager of every high-rise residential building be required by law to prepare personal emergency evacuation plans (PEEPs) for all residents whose ability to self-evacuate may be compromised (such as persons with reduced mobility or cognition).

Recommendation 33.22f

That the owner and manager of every high-rise residential building be required by law to include up-to-date information about persons with reduced mobility and their associated personal emergency evacuation plans (PEEPs) in the premises information box.

Recommendation 33.22g

That all fire and rescue services be equipped with smoke hoods to assist in the evacuation of occupants through smoke-filled exit routes.

Recommendation 33.27

That in all high-rise buildings floor numbers be clearly marked on each landing within the stairways and in a prominent place in all lobbies in such a way as to be visible both in normal conditions and in low lighting or smoky conditions.

Recommendation 33.28

That the owner and manager of every residential building containing separate dwellings (whether or not it is a high-rise building) be required by law to provide fire safety instructions (including instructions for evacuation) in a form that the occupants of the building can reasonably be expected to understand, taking into account the nature of the building and their knowledge of the occupants.

Recommendation 33.29a

That the owner and manager of every residential building containing separate dwellings (whether or not they are high-rise buildings) carry out an urgent inspection of all fire doors to ensure that they comply with applicable legislative standards.

Recommendation 33.29b

That the owner and manager of every residential building containing separate dwellings (whether or not they are high-rise buildings) be required by law to carry out checks at not less than three-monthly intervals to ensure that all fire doors are fitted with effective self-closing devices in working order.

Recommendation 33.30

That all those who have responsibility in whatever capacity for the condition of the entrance doors to individual flats in high-rise residential buildings, whose external walls incorporate unsafe cladding, be required by law to ensure that such doors comply with current standards.

Recommendation 33.31a

That the Joint Doctrine be amended to make it clear that each emergency service must communicate the declaration of a Major Incident to all other Category 1 Responders as soon as possible.

Recommendation 33.31b

That the Joint Doctrine be amended to make it clear that on the declaration of a Major Incident, clear lines of communication must be established as soon as possible between the control rooms of the individual emergency services.

Recommendation 33.31c

That the Joint Doctrine be amended to make it clear that a single point of contact should be designated within each control room to facilitate such communication.

Recommendation 33.31d

That the Joint Doctrine be amended to make it clear that a 'METHANE' message should be sent as soon as possible by the emergency service declaring a Major Incident.

Recommendation 33.32

That steps be taken to investigate the compatibility of the London Fire Brigade systems with those of the Metropolitan Police Service and the London Ambulance Service with a view to enabling all three emergency services' systems to read each other's messages.

Recommendation 33.33

That steps be taken to ensure that the airborne datalink system on every National Police Air Service helicopter observing an incident which involves one of the other emergency services defaults to the National Emergency Service user encryption.

Recommendation 33.34

That the London Fire Brigade, the Metropolitan Police Service, the London Ambulance Service and the London local authorities all investigate ways of improving the collection of information about survivors and making it available more rapidly to those wishing to make contact with them.

Many of the recommendations made in the Phase One Report have been implemented.

What were the recommendations in the Inquiry Phase Two Report?

The Phase Two Report was published in September 2024.

The Inquiry Chair, Sir Martin Moore-Bick, set out recommendations for the construction industry, architecture and government in the Phase Two Report in order to avoid another disaster of Grenfell scale.

Architects: the Inquiry recommended improvements to architectural education, training and the role of the Principal Designer.

The report said

> We recognise that both the Architects Registration Board (ARB) and the Royal Institute of British Architects (RIBA) have taken steps since the Grenfell Tower fire to improve the education and training of architects. We recommend that they should review the changes already made to ensure that they are sufficient in the light of our findings.
>
> We also recommend that it be made a statutory requirement that an application for building control approval in relation to the construction or refurbishment of a higher-risk building be supported by a statement from a senior manager of the principal designer under the Building Safety Act 2022 that all reasonable steps have been taken to ensure that on completion the building as designed will be as safe as is required by the Building Regulations.

Creation of a single independent body headed by a person 'construction regulator'

The Inquiry called for a single body reporting to a single Secretary of State to bring about 'a much-needed change on the culture of the construction industry'. Its role would include oversight of certification, compliance, regulation of building control and the licensing of contractors working on high-rise buildings.

This includes changes to the Building Safety Act 2022.

For the purposes of the Building Safety Act 2022, the report called on the government to urgently review the definition of a higher-risk building: 'we do not think that to define a building as "higher-risk" by reference only to its height is satisfactory, being essentially arbitrary in nature' and the meaning should be 'more relevant' by reflecting 'the nature of its [the building's] use'.

Combining government fire safety responsibilities under a new Secretary of State

The 'fragmentation of responsibility' for regulating the construction industry is currently duplicated and confusing in the range of government departments responsible for matters affecting fire safety.

The report therefore recommended bringing together fire safety responsibilities currently spread between the Ministry of Housing, Communities and Local Government, the Home

Office and the Department of Trade into 'one department under a single Secretary of State'. It added that this Secretary of State should appoint a chief construction adviser.

Review of Approved Document B

The Inquiry panel said it did 'not think that Approved Document B provides the information needed to design buildings that are safe in fire' and has demanded that it is reviewed 'as a matter of urgency'.

The report also recommended that 'it be made a statutory requirement that a fire safety strategy produced by a fire engineer to be submitted with building control applications' for construction or refurbishment of higher-risk buildings and for a later review stage.

New licensing scheme for contractors

The report highlighted how design and build contractor Rydon failed to make it clear which subcontractor was 'responsible for particular aspects of the design' and had not taken an 'active interest in fire safety'. It reads: 'We are not the first to conclude that the construction industry as a whole needs to become technically more competent and less willing to sacrifice quality to speed and cost.'

The Inquiry recommended 'a licensing scheme operated by the [new] construction regulator' to issue licenses for Principal Contractors working on construction or refurbishment of higher-risk buildings and that a senior manager or director of the Principal Contractor must support the application for building control and agree that all reasonable care has been taken to ensure that on completion and handover the building meets Building Regulations.

National building control authority

Recommendation was made for the building control function to be carried out by a national authority so as to avoid potential conflicts of interest with any local regulatory organisation.

Regulation of fire engineers with new register

The Inquiry pointed out that, currently, the term 'fire engineer' does not denote any formal qualification and as a result it is possible for anybody to practise as a fire engineer.

The report called for the profession of fire engineer to be formally recognised and protected by law with an independent regulator, fulfilling a similar role to the Architects Registration Board. It also recommended that the government convene a group of practitioner and academic fire engineers, and other professionals, to produce an authoritative statement of the knowledge and skills to be expected of a competent fire engineer.

Competency of fire risk assessors

The Inquiry recommended that the government establish a system of mandatory accreditation to certify the competence of fire risk assessors by setting standards for qualification and continuing professional development and such other measures as may be considered necessary or desirable: 'We think it necessary for an accreditation system to be mandatory in order to ensure the competence of all those who offer their services as fire risk assessors.'

Government response to recommendations made by various bodies, coroners and so forth

The report recommended that it is made a legal requirement for the government to maintain a record of recommendations made by select committees, coroners and public inquiries, and for the government to explain why it does not agree with recommendations, with parliamentary scrutiny.

Establishing a national College of Fire and Rescue

After finding major issues with standards in the fire service, the Inquiry recommended the creation of a College of Fire and Rescue that would carry out practical training, lectures and research and set national standards.

The Inquiry panel made further recommendations concerning local authorities, gas pipe safety, local resilience forums, response and recovery, deployment of firefighters, vulnerable people, gold command communications, hydrant knowledge and training, and a number of ancillary subjects.

Will anything change as a result of these recommendations?

In February 2025, the UK government issued a statement in the House of Commons that it has accepted in full 49 of the 58 recommendations and accepted the remaining nine in principle.

Some of the recommendations have been incorporated into the Building Safety Act 2022.

Key findings of the Grenfell Tower Inquiry Phase Two Report

- Systematic dishonesty about panels and insulation. Testing processes and misrepresented test data that was intended to mislead the market was provided by producers and suppliers of panel and insulation products.
- Regulation at the time was seriously defective. The system of regulating the construction and refurbishment of high-rise residential buildings in place at that time was defective in a number of respects.
- Conflicts of interest were inherent in the system. The report revealed an inappropriate relationship between approved building inspectors and those they were inspecting. Approved inspectors had a commercial interest in acquiring and retaining customers, which conflicted with the performance of their role as guardians of the public interest and, as a result, the inspectors sacrificed rigorous application of principle to its commercial interests.
- The government knew of the risks for years. Twenty-six years prior to Grenfell in 1991, an 11-storey tower block in Merseyside suffered a serious fire that involved combustible cladding panels and insulation. The department was aware of the

risks but failed to act. In 1999, the Environment and Transport Select Committee warned the government not to wait until a deadly fire occurred for steps to be taken to minimise the risks posed by some external cladding systems.

- Uncaring attitudes and bullying. Relations between the Tenant Management Organisation (TMO) responsible for running services at Grenfell Tower and its residents were found to be increasingly characterised by distrust, dislike, personal antagonism and anger. Some occupants regarded the TMO as an uncaring and bullying overlord that belittled and marginalised them. The report found that there was a toxic atmosphere fuelled by mistrust on both sides, but there was a serious failure by the TMO in allowing the relationship to deteriorate to such an extent. The TMO and the Royal Borough of Kensington and Chelsea were jointly responsible for managing fire safety at Grenfell Tower but the years between 2009 and 2017 were marked by a persistent indifference to fire safety.
- The need for better regulation. One of the foremost recommendations in the report was to establish a single independent body to regulate the industry and drive change, with this body reporting to a single government Secretary of State. Its functions would include regulating construction products, testing and certification, licensing of contractors to work on higher-risk buildings, carrying out research, and accrediting fire risk assessors. It suggested that the Secretary of State this body would report to should have a chief construction adviser. The report also called for the profession of 'fire engineer', which currently requires no formal qualification to practise, to be formally recognised, and for the title and function to be protected by statute.

The Building Safety Act 2022: Questions and Answers: A practical approach to design and management of higher-risk buildings

Pat Perry
ISBN 978-1-83662-467-7
https://doi.org/10.1108/978-1-83662-464-620251003

Chapter 3
Understanding the Building Safety Act 2022

Does the Building Safety Act apply to all of the UK?

No. The Act applies to England although selected provisions will apply to the devolved nations.

Wales has adopted in principle the Building Safety Act and will introduce specific measures via regulations in due course.

Scotland will be taking a different route to improving building safety in respect of high-rise buildings and plans to amend current legislation, especially via the building code and regulations framework. It has recently introduced the Housing (Cladding Remediation) (Scotland) Act 2024 to address the issue of external cladding and remedial works.

Northern Ireland would appear to be adopting a similar route to Scotland and will be amending existing legislation.

What is the Building Safety Act?

The Building Safety Act takes forward the government's commitment to fundamental reform of the building safety system. It introduces a new regulatory regime for high-risk residential buildings, now defined as 'higher-risk buildings' and buildings 'in scope' – that is, lower rise buildings that demonstrate a higher fire safety risk.

The Explanatory Notes to the Act provide a proposed definition of higher-risk buildings as being

> the 'higher-risk buildings', which will be regulated by the new occupation regime, as at least 18 metres in height or having at least 7 storeys and containing at least two residential units. Residential units are either a dwelling or any other unit of living accommodation, for example student accommodation where basic amenities, such as cooking facilities, a toilet and personal washing facilities are shared with others in the building.

Houses, flats, serviced apartments, supported accommodation and student accommodation facilities are all expected to be within the scope of higher-risk buildings.

Residential care homes, prisons, hotels, hospitals and hospices are likely to be excluded from the provisions of the new Act but may be included later under separate regulations.

The Secretary of State is authorised to update the list of higher-risk buildings, based on emerging risk evidence.

Part 2 of the Act establishes a new national **Building Safety Regulator**, which will sit in the Health and Safety Executive and report to the Secretary of State. The Regulator will be responsible for implementing the new regulatory regime, overseeing the registration and inspection of higher-risk buildings, and has express powers to authorise remedial works, stop non-compliant projects, appoint special measures for failing projects and order the replacement of key fire safety officers. The Regulator will establish and consult with three advisory committees on building, industry competence and resident representation while developing the new building safety regime.

Part 3 of the Act empowers the Secretary of State to impose **competency requirements** on duty holders working on higher-risk buildings. Those appointing duty holders will also be required to ensure all appointees and prescribed persons meet competency requirements and provide documents demonstrating competency. Full details of duty holder responsibilities and the new gateway system for approving the design, construction and occupation of in-scope buildings will be covered in secondary legislation.

Part 4 of the Act establishes the role of the **Accountable Person** who will be legally responsible for the safety of higher-risk buildings. Accountable Persons may be individuals or corporate entities and will hold either a legal estate in possession of the common parts of the building or a relevant repairing obligation in respect of the common parts (meaning that management companies may also be Accountable Persons).

Accountable Persons have ongoing obligations to assess and prevent fire safety risks, update prescribed building safety information (referred to as 'the golden thread'), register higher-risk buildings and apply for a Building Assessment Certificate before the building can be occupied.

Accountable Persons must proactively engage with residents, provide them with key building information and develop a residents' engagement strategy.

The Act updates the Landlord and Tenant Act 1985, placing express duties on residents of higher-risk buildings to keep their property in good repair, cooperate with Responsible Persons to fulfil building safety duties and not hinder or damage fire safety works or equipment.

Accountable Persons will have rights to enter residents' dwellings and issue enforcement notices, which shall be enforceable via the courts. The Act imports a new regime for charging tenants building safety costs incurred carrying out building safety work. However, landlords will not be able to recharge residents for work that are the subject of enforcement proceedings by the Regulator or as a result of the Accountable Person's negligence or breach of contract.

The Act will impose implied terms on lease agreements covered by the 1985 Act to incorporate this new regime.

The Act provides details about liability and sanctions for breaches of the building safety regime. Duty holders who fail to meet key building safety obligations will be guilty of criminal offences.

Where a duty holder is a corporate entity, individuals within that organisation may also be prosecuted where the breach was committed with their consent or connivance or because of their neglect.

Accountable Persons who fail to meet key obligations will also be guilty of a criminal offence, punishable by imprisonment for up to 2 years and/or a fine.

The Act also amends the Building Act 1984 to allow prosecutions for breaches of the Building Regulations for up to 10 years. However, the Act does not anticipate duty holders being placed under statutory duties or allowing individuals to bring actions in civil court against duty holders.

Part 5 of the Act requires the Secretary of State to establish a scheme for all developers to allow complaints to be made to a newly established **New Homes Ombudsman.**

The Secretary of State will issue guidance about the adequacy of insurance schemes for higher-risk buildings and building safety works.

The Secretary of State will also establish a new regulatory system for construction products marketed in the UK, identifying 'designated standards' and 'safety-critical products' that may be withdrawn from the market if they present a risk.

The Act also amends the Architects Act 1997 to allow the Architects Registration Board to monitor the competence of architects in line with the new regime.

The Building Safety Act will also amend the Regulatory Reform (Fire Safety) Order 2005.

Part 6 of the Act covers miscellaneous matters such as liability of officers of body corporate and so forth, review of regulatory regime, financial provisions, Crown application, application to Parliament, power of Secretary of State to make consequential provision, power of Welsh Ministers to make consequential provision, regulations, extent, commencement and transitional provision and short title.

Why was the Building Safety Act 2022 needed?

The following is an extract from the government's response to the Grenfell Tower Inquiry report.

> The Rt Hon Angela Rayner MP: Deputy Prime Minister and Secretary of State for Housing, Communities and Local Government
>
> The fire at Grenfell Tower on 14 June 2017 that took 72 lives was a national tragedy. It was also a personal tragedy for those bereaved, the survivors who lost their homes, and the community of which they were all a part.

> Sir Martin Moore-Bick and his team have uncovered and faithfully recorded evidence of failure after failure, year after year. The government and regulators failed to protect people. The Royal Borough of Kensington and Chelsea and its Tenant Management Organisation failed to listen to residents. The London Fire Brigade was unprepared for a fire like Grenfell. The companies that refurbished Grenfell Tower put profits ahead of safety – and, worse, some concealed how flammable their cladding and insulation products were, despite knowing they would be used on people's homes.
>
> The Inquiry has laid bare the truth in painstaking detail, amplifying the demands and the need for justice. I understand that for many this means criminal charges. The government fully supports the Metropolitan Police in its investigation.
>
> We must face the stark reality that the failings that led to the events on 14 June 2017 were not unique to Grenfell Tower. Thousands of people are still living in buildings with combustible cladding across the country. The impact on their lives is completely intolerable and the speed at which this has been addressed has been too slow.
>
> I have heard from the community the need for government to respond fully to the Inquiry's recommendations and to take further action to instil long-lasting change. The government accepts all the Inquiry's findings and the problems that it highlighted need to be fixed. We are committed to taking forward the work required to address the Inquiry's recommendations, and indeed to going further: we must embrace not only the specific findings, but their spirit. Anyone who lives in an unsafe home is one person too many. This must be the moment that we refuse to accept anything less than safe, quality homes for everyone.
>
> We must build a regulatory system that everyone trusts to deliver this. As well as responding to each of the Inquiry's important recommendations, this publication lays the foundations for long-term reform that will create better, safer buildings. Today, we set out a vision to reset the relationship between the government, the public, the housing sector, our fire and rescue authorities and their services.

The regulatory system in place before the Grenfell Tower fire happened failed completely in parts and was found to be inadequate in others. Reform was essential if a catastrophe such as the Grenfell Tower fire was to be avoided in the future.

The Building Safety Act 2022 has been enacted in order to address some of those failures.

What is the role of the Building Safety Regulator?

The Building Safety Regulator will improve the safety and standard of buildings through three critical functions

- leading the delivery of the new regulatory regime for high-rise and other buildings in scope
- overseeing the safety and performance of all buildings
- promoting the competence and organisational capability of professionals, tradespeople and building control professionals working on all buildings.

The Building Safety Regulator will work closely with local authorities and fire and rescue authorities, bringing in external technical experts as necessary, to make sure the right specialists and regulators work effectively together to regulate high-rise and other buildings in scope; create expert committees, including a statutory residents' panel, which will ensure that residents have a strong voice in the development of policy and strategy within the Building Safety Regulator; and empower the Building Safety Regulator to charge fees to recover costs from regulated parties.

What are the duties of the Building Safety Regulator?

There are three main duties allocated to the Building Safety Regulator, namely

- to oversee the safety of people in relation to risks arising from buildings and to improve the standards of all buildings
- to directly regulate the safety of higher-risk buildings during construction and after occupation
- to improve the competence of all those involved in the construction and management of higher-risk buildings.

The Building Safety Regulator will be the Building Control authority for higher-risk buildings.

Other duties and responsibilities of the Building Regulator will be

- the ability to publish a new Code of Conduct for building inspectors
- to carry out misconduct investigations
- to issue stop notices
- to levy unlimited civil fines and commence criminal prosecutions that attract prison sentences of up to 2 years
- to extend its prosecutorial powers to directors, managers and officers of companies that have committed an offence with their consent or connivance, and to extend time limits for prosecution to 10 years for certain notices issued under the Building Act 1984
- to appoint a Special Measures Manager where there are serious failures on the part of Accountable Persons or Building Safety Managers.

A new role of 'Accountable Person' is to be created under the Building Safety Act. Who is this and what are their duties?

The Accountable Person is the entity responsible for meeting the statutory obligations for occupied higher-risk buildings. The Accountable Person will usually be the person who either has the legal estate in possession of, or is under a relevant repairing obligation for, any part of the common parts of the building.

An Accountable Person is an organisation or individual who owns or has a legal obligation to repair any common parts of the building.

Common parts are used by residents, such as

- the structure and exterior of the building
- corridors
- lobbies
- staircases.

An Accountable Person can be a

- freeholder or estate owner
- landlord
- management company
- resident management company
- right to manage company
- commonhold association.

Where there are two or more persons with legal interest in a building in scope, the Accountable Person will be the one responsible for the repair and maintenance of any common parts of the building.

Where there are two or more Accountable Persons for a building, the person/organisation responsible for the exterior and structure of the building will be designated 'Principal Accountable Person'.

Example

In a building with a single Accountable Person, that Accountable Person will be responsible for all the Accountable Person duties in Part 4 of the Building Safety Act.

In a building where there is a freeholder which retains repairing obligations for the exterior and structure of the property, and there is a superior leaseholder which retains repairing obligations for all the other 'inner' common parts, both entities will be Accountable Persons and the freeholder will be the Principal Accountable Person.

In a building where there is a freeholder which does not retain repairing obligations, a superior leaseholder with repairing obligations for the exterior and structure of the property and a management company where every lease in the property states that the management company is responsible for the remainder of the common parts, the superior leaseholder and the management company will be Accountable Persons. The superior leaseholder will be the Principal Accountable Person.

(Example taken from government Explanatory Notes on the Building Safety Act July 2021)

This is the legal person who under the new regime will be responsible for the fire and structural safety of a 'higher-risk' building.

The Accountable Person will be responsible for assessing building safety risks on an ongoing basis, and for taking all reasonable steps to manage those risks by preparing a Safety Case Report, which must be submitted to the Building Safety Regulator.

New criminal sanctions (including imprisonment of up to 2 years and unlimited fines) will apply to the Accountable Person. Criminal liability could also be attached to individual directors, managers and members of the Accountable Person, if it is a body corporate, where there is consent, connivance or neglect by the individual.

The Accountable Person could be the building owner, the managing agents or any individual or corporate entity that has repairing obligations for the building.

An Accountable Person's responsibilities include the following.

- Registering a building with the Building Safety Regulator.
- Applying for a Building Assessment Certificate, which is issued when the Building Safety Regulator is satisfied the Accountable Person is complying with their duties under the Act.
- Assessing (and revising as necessary) building safety risks and taking all reasonable steps to prevent the occurrence, and control the impact, of a major incident arising from building safety risks in or around the building.
- Preparing (and revising as necessary) a Safety Case for a building that contains their assessment of the building safety risks and the steps that have been taken to prevent a major incident.
- Keeping information on a building and ensuring it is up to date.
- Developing systems for investigating complaints and for mandatory occurrence reporting.

The Building Safety Act 2022 refers to a Principal Accountable Person. Who is this and what duties do they have?

Each building must have one clearly identifiable Accountable Person, known as the Principal Accountable Person.

The Act provides that where there is only one Accountable Person, that person will automatically become the Principal Accountable Person.

If a building has different Accountable Persons for different parts of the building, the Act states that the Principal Accountable Person will be the one responsible for the structure and exterior of the building.

A Principal Accountable Person would be subject to these additional ongoing duties and obligations.

- Registering the building – details of how to apply for registration and what information is required is to be set out in secondary legislation, although this requirement will apply to both new and existing buildings.
- Applying for a Building Assessment Certificate.
- Displaying the Building Assessment Certificate in a prominent location – it would be a criminal offence not to.
- Preparing a Safety Case Report – this is a report that details all of the building safety risks identified by the Accountable Persons in respect of their part and also sets out a brief description of what action those Accountable Persons are taking in relation to those risks.
- Establishing and operating the mandatory occurrence reporting system – this ties in with the ongoing duty that Accountable Persons have to report prescribed information to the Building Safety Regulator. It will be the Principal Accountable Person's responsibility to establish and create a framework through which the Accountable Persons can make these reports.
- Responsibility for preparing a residents' engagement strategy and establishing a complaints procedure.

The Principal Accountable Person is usually an organisation, like a commonhold association, local authority or social housing provider.

In some circumstances, an individual can be the Principal Accountable Person. For example, the individual is

- the owner of the building
- named on the leasehold as responsible for the maintenance of common parts.

When there are multiple Accountable Persons, then whoever owns or has a legal obligation to repair the structure and exterior of the building is the Principal Accountable Person.

Accountable Persons and the Principal Accountable Person can be accountable for the fire and structural safety risks of more than one building.

Accountable Persons and the Principal Accountable Person cannot delegate their legal obligations to others.

Accountable Persons and the Principal Accountable Person can employ an organisation or individual, like a managing agent, to carry out duties on their behalf. However, the accountability for making sure those duties are carried out and the liability for a building's safety remains with the Accountable Persons and Principal Accountable Person.

Organisations who have responsibility as Responsible Persons or Principal Responsible Persons should designate a named individual as the point of contact for the organisation.

The single point of contact can also be a third party separate to the organisation, such as a management company. This individual should have authority or duties relating to the safety of the building, but this does not make them or the third party the Principal Accountable Person. It is the organisation that is the Principal Accountable Person.

The Principal Accountable Person must give written consent to the third party. If this consent stops, the Principal Accountable Person must update the single point of contact with the Building Safety Regulator.

What is a 'higher-risk building' under the Building Safety Act?

Exact wording in the Act states that

> "higher-risk building" means a building in England that—
>
> (a) is at least 18 metres in height or has at least 7 storeys, and
> (b) contains at least 2 residential units.

The definition in the Act of residential unit is

> "residential unit" means—
>
> (a) a dwelling, or
> (b) any other unit of living accommodation.

Therefore, in addition to residential buildings, care homes and hospitals can also be considered higher-risk buildings if they are higher than 18 m or have at least seven storeys and so would fall under the new regulatory regime.

The Secretary of State has issued The Higher-Risk Buildings (Descriptions and Supplementary Provisions) Regulations 2023. The Regulations are divided into definitions for buildings during design and construction and those during occupation.

A 'higher-risk building' during design and construction is defined as a building that is at least 18 m in height, or has at least seven storeys, and contains at least two residential units; or is a care home; or is a hospital.

A 'higher-risk building' during occupation is one that is at least 18 m in height, or has at least seven storeys, and contains at least two occupied residential units.

Buildings will be excluded from the definition of 'higher-risk buildings' during occupation, and do not need to be registered, if they are used in their entirety as one of the below

- hospital
- care home
- secure residential institution
- hotel
- military barracks and living accommodation for military personnel
- prison.

All higher-risk buildings must be registered with the Building Safety Regulator either before occupation of a new build or by 30 September 2023 if an existing building in occupation.

How is a building storey defined?

The general definition is as follows.

A storey is generally considered the space between two consecutive floors or a floor and a roof.

Storeys are counted from the lowest external ground level.

Storeys below ground level are generally not counted towards the total number of storeys.

A mezzanine floor is considered a storey if its internal floor area is at least half of the largest storey's internal floor area.

The question of whether roof terraces and similar spaces should be counted as storeys has been debated, with some cases suggesting they should be counted if they are usable.

Building Safety Act 2022 section 118 defines height of buildings and number of storeys in the following.

(1) This section applies for the purpose of section 117 (meaning of relevant building).
(2) The height of a building is to be measured from ground level to the finished surface of the floor of the top storey of the building (ignoring any storey which is a roof-top machinery or plant area or consists exclusively of machinery or plant rooms).
(3) When determining the number of storeys in a building—

(a) any storey below ground level is to be disregarded;
(b) any mezzanine floor is to be regarded as a storey if its internal floor area is at least half of the internal floor area of the largest storey in the building which is not below ground level.

(4) In subsection (2) "ground level", in relation to a building, means—

(a) the level of the surface of the ground immediately adjacent to the building, or
(b) where the level of the surface of the ground on which the building is situated is not uniform, the level of the lowest part of the surface of the ground immediately adjacent to it.

(5) For the purposes of subsection (3) a storey is "below ground level" if any part of the finished surface of the ceiling of the storey is below the level of the surface of the ground immediately adjacent to that part of the building."

How does the Building Safety Act 2022 influence fire safety matters?

The Building Safety Act will amend the Regulatory Reform (Fire Safety) Order 2005 to require that all Responsible Persons will record their Fire Risk Assessments and fire safety arrangements and appoint only competent persons to undertake those assessments.

Responsible Persons will also need to work with other Responsible Persons in the same premises. In the case of residential higher-risk buildings, Responsible Persons will need to co-operate with the Accountable Person.

The Building Safety Act 2022 will also strengthen the role of the Principal Designer under the Construction (Design and Management) (CDM) Regulations 2015 so that as well as construction site safety duties, Principal Designers will now be expected to plan, manage and monitor design work to ensure compliance with the Building Regulations – and maintain a 'golden thread' of data on design decisions for fire and structural safety.

What are building safety risks as defined in the Building Safety Act?

A 'building safety risk' is defined as any risk to the safety of persons in or about a building arising from fire, structural failure, or any other matter prescribed in Regulations (section 62).

The clause is drafted to capture risks inside and outside the building such as fire spreading on the external wall of a building but is limited to risks that are associated with the building and/or arising from the building itself. Risks are also limited to those affecting people in and about the building.

The explanatory notes to the Act state that the scope is only intended to encompass major and rapid onset events and that the government 'has no current plans for the higher-risk building regime to extend beyond the risks of fire and structural integrity'.

However, the Secretary of State can request advice from the Building Safety Regulator about new and emerging risks and may introduce additional definitions for building risk.

Statutory guidance may also be issued to help determine building risk and may include electrical and gas failures, integrity of windows and glazing due to high winds and so forth.

The Higher-Risk Buildings (Management of Safety Risks etc) (England) Regulations 2023 give clarity to how building risk is to be interpreted.

The principles prescribed in Regulation 4 for the purposes of section 84(3) of the Building Safety Act 2022 are to

- avoid building safety risk
- evaluate building safety risks that cannot be avoided, including identifying the proportionate measures required to reduce, address and mitigate the building safety risks
- combat building safety risks at source by introducing proportionate measures to reduce, address and mitigate that risk at the earliest opportunity
- ensure suitable and proportionate systems are in place for the effective inspection, testing and maintenance of the efficacy of measures taken
- give collective protective measures priority over individual protective measures

- adapt to technical progress
- where reasonable to do so, replace the dangerous with the non-dangerous or less dangerous.

Who is responsible for managing the building risks associated with a higher-risk building?

The Accountable Person is responsible.

Every Accountable Person is under an ongoing obligation to carry out an assessment of the building safety risks in/of/arising from, and relevant to, the part of the higher-risk building for which they are responsible. They must take all reasonable steps and actions necessary to manage the building safety risks, preventing them from materialising in their building and to mitigate the severity of such an incident if it does occur.

What is the Safety Case?

The Safety Case Report should refer to the totality of the building safety information and include all supporting evidence identifying how fire and structural risks are being managed by the Principal/Accountable Person. This should be followed by a written explanation and justification of the approach being taken to manage risks, referencing the supporting evidence in the Safety Case.

The Safety Case Report is submitted by the Accountable Person to the Building Safety Regulator. In submitting a Safety Case Report for review, the Accountable Person will make a case to the Building Safety Regulator that the fire and structural safety risks in the building have been assessed, and adequate safety measures have been identified and put in place to manage the risks appropriately.

Key to the process of putting the Safety Case Report together will be the established risk assessment principles of

- identifying the hazards
- deciding who might be harmed and how
- evaluating the risks associated with these hazards
- deciding on the necessary control and mitigation measures
- recording those findings and implementing them
- evaluating and monitoring on an ongoing basis.

The Safety Case Report will include all relevant information on the structure and materials used in the building and the strategies adopted during the design and construction of the building to minimise risks from catastrophic events.

The Safety Case Report will include a comprehensive Fire Risk Assessment report and emergency plan as required by the Regulatory Reform (Fire Safety) Order 2005.

The Building Safety Regulator will assess the Safety Case Report and will either accept the finding in the report or will require amendments and improvements to the safety regime proposed.

What is the 'golden thread of information'?

Accurate and up-to-date information about the design, construction and ongoing maintenance of buildings is crucial to ensuring building safety. The golden thread, which will operate from design stage through to occupation of buildings in scope, is intended to encourage openness, transparency and accountability.

The intention of the clauses in the Act is to ensure that the right people have the right information at the right time to ensure buildings are safe and building safety risks are managed throughout the building's life cycle.

This information will be held digitally and will ensure that the original design intent and any subsequent changes to the building are captured, preserved and used to support safety improvements.

It is recommended that Building Information Modelling (BIM) is used to collate and share the information.

The Construction (Design and Management) Regulations 2015 require that a Health and Safety File is produced by the Principal Designer for all projects where there is more than one contractor. The Health and Safety File will contain information relevant to health and safety and can be provided in either digital or hard copy formats. Ideally, the Health and Safety File can be extended to include all the relevant information about the construction project.

Certain information from the golden thread will need to be submitted to the Building Safety Regulator at the beginning of the project, namely

- full plans, including a detailed specification in relation to how fire and structural safety risks will be managed
- a three-dimensional (3D) digital model of the building 'as planned', which must also show the construction products to be used
- a fire statement, addressing fire service access and water availability. This will ultimately be developed into a Fire and Emergency File, building on the fire statement and updated throughout the construction phase
- a Construction Control Plan setting out how compliance with the Building Regulations will be achieved and how changes will be controlled and recorded.

Major changes affecting safety will need to be agreed with the Principal Designer and Client and approved by the Building Safety Regulator in advance.

Additional information will need to be submitted to the Building Safety Regulator before the building can be released for occupancy, namely

- handover of building safety information on the final, 'as-built' building
- Safety Case Report for the building, identifying hazards, describing how risks are controlled and detailing safety management systems, including emergency response procedures

- residents' engagement strategy setting out how information will be provided to residents
- complaints procedures and wider engagement initiatives.

The Safety Case Report must be reviewed and re-submitted to the regulator every 5 years and also if the building undergoes major refurbishment.

Under the new regime, a Building Assessment Certificate will be required before a building can be occupied. The regulator will not grant a Building Assessment Certificate unless all the relevant information has been submitted and approved.

What is the Assessment Certificate?

Assessment Certificate will be a legal document issued by the Building Safety Regulator that confirms that the building has been constructed in accordance with all the statutory provisions, and that procedures are in place to manage all major risks that could affect fire safety, structural integrity and any other prescribed matter of the building.

It will be issued when all the relevant information, including the Safety Case Report, has been submitted by the Approved Person.

The Building Safety Regulator can issue the Building Assessment Certificate with conditions.

Residents will not be permitted to occupy a new development without an official Building Assessment Certificate.

A copy of the certificate will need to be displayed in all common parts of the building so that residents know that the building has been designed with safety in mind and have knowledge of the names and contact details of the Accountable Person.

What are the 'gateways' referred to in the Building Safety Act?

The Building Safety Act talks about three gateways during the planning, design and construction phases of the building.

Each of these gateways has certain requirements that would need to be met through the cooperation of various duty holders.

The first is the planning gateway that focuses on the planning permission process.

Gateway two must be met before construction begins, whereas the third gateway needs to be met before the occupation of the building.

A similar process has been proposed for major building refurbishments and for buildings that switch from commercial to residential usage. At each gateway, the golden thread of information will need to be updated to get to the next stage.

Gateway one: before planning permission is granted

The first gateway is the least onerous of the three new gateways. Its purpose is to ensure that there is consultation with the local fire authority at the earliest possible stage.

Gateway one will apply only to new high-rise residential buildings of 30 m or more.

It may also be extended to other developments in the 'near vicinity' of these buildings, depending on the outcome of the consultation.

The main features of gateway one are

- making local fire and rescue authorities 'statutory consultees' during the planning process, requiring the local planning authority to consult with them before granting planning permission; and
- requiring the Client to submit a fire statement covering fire service vehicle access and access to water supplies with the planning application.

Gateway two: before construction begins

Gateway two will apply to new multi-occupied buildings of 18 m or more, and to significant refurbishment of existing buildings in scope of the new Act. The new requirements supplement the current 'full plans building application' stage under the Building Regulations 2010, and encourage very early dialogue between the regulator and duty holder about risks and how they are being managed.

Duty holders will need to prepare and submit a number of pieces of the 'golden thread' of information about the building

- full plans, including a detailed specification in relation to how fire and structural safety risks will be managed
- 3D digital model of the building 'as planned' – this must also show the construction products to be used
- a Fire and Emergency File, building on the fire statement from gateway one and updated throughout construction
- a Construction Control Plan setting out how compliance with the Building Regulations will be achieved and how changes will be controlled and recorded.

Major changes affecting safety will need to be agreed with the Principal Designer and Client and approved by the Building Safety Regulator in advance.

A phased approach will be possible for complex development, which would result in multiple gateway two approvals for different parts of the development.

An inspection regime must also be agreed with the new regulator as part of gateway two. The regulator will have the power to require demolition of non-compliant building work, or for that work to be laid open for inspection for compliance.

Gateway three: before occupation

The third gateway is designed to ensure an adequate handover of the building from those involved in its construction to those who will be responsible for managing and maintaining it during occupation.

Gateway three will be introduced on a phased basis for all existing buildings within the scope of the new regulatory regime. Existing building owners/managers will need to carry out significant work to recreate the golden thread of information for these buildings.

The following will need to be in place before gateway three approval will be granted.

- A declaration from the Principal Contractor and the Principal Designer confirming the building complies with the Building Regulations.
- Handover of building safety information on the final, as-built building plans.
- Registration of the building.
- Registration of the Accountable Person responsible for building safety.
- A Safety Case for the building identifying hazards, describing how risks are controlled and detailing safety management systems including emergency response procedures.

Once the regulator is satisfied, it will issue a Building Assessment Certificate permitting occupation. The regulator may attach conditions to the certificate. The certificate and Safety Case will be subject to review every 5 years.

What is mandatory occurrence reporting under the new Building Safety Act?

The Principal Accountable Person has a duty to establish and operate an effective mandatory occurrence reporting system and a further duty is placed on Accountable Persons to ensure that reportable information related to the safety of an occupied building is given to the Building Safety Regulator in the manner specified by the Building Safety Regulator. The system so adopted will be assessed as part of the Building Assessment Certificate application.

Basically, the Principal Accountable Person must have in place a system to report particularly dangerous safety-related events to the Building Safety Regulator.

As an example – an outbreak of fire in the refuse chutes/room that failed to shut down the ventilation system or automatically close fire doors due to faults must be reported as serious consequences to life safety could have ensued.

The Building Safety Regulator will impose remedial actions and may check that works have been carried out.

Information has to be kept in relation to higher-risk buildings. What does this mean?

The Accountable Person must keep and maintain certain required information about their buildings and ensure that it is kept as up to date as possible. This information is part of the 'golden thread' of information.

The Building Safety Act does not define exactly what will be prescribed information as this is be set out in additional regulations, The Higher Risk Buildings (Key Building Information etc) (England) Regulations 2023.

Information could include

- details of the fire alarm system
- means of escape details
- fire stopping information
- the Fire Risk Assessment
- emergency plans for people with disabilities
- maintenance schedules for all fire protection and preventative measures
- details of the local Fire and Rescue Authority
- schedule of building materials used within the building.

The Accountable Person will be under an obligation to share the 'prescribed information' with others who may need to know about the building – for example, contractors who are carrying out refurbishment work, residents and residents' groups, the Building Safety Regulator and the local Fire and Rescue service.

Who will be duty holders under the new Act?

Duty holders will be

- the Accountable Person or Principal Accountable Person
- the Client
- the Principal Designer
- the Principal Contractor
- designers
- contractors.

The duties of the Client will be

- establishing a culture of compliance on the project
- project arrangements: making suitable arrangements for managing the building work to deliver compliance with the Building Regulations, including the allocation of sufficient time, resources and prioritisation
- key appointments: appointing a competent Principal Designer and Principal Contractor, with named individuals registered as competent with the new Building Safety Regulator
- compliance: taking reasonable steps to ensure that the Principal Designer and Principal Contractor comply with their building safety obligations
- golden thread: establishing appropriate information management arrangements for the building
- gateways: ensuring that regulatory requirements to get approval from the new Building Safety Regulator at various stages of the project are met
- handover: ensuring an appropriate pre-occupation handover takes place.

The duties of the Principal Designer will be

- leading the design: planning, monitoring and managing the pre-construction phase and coordinating building safety to ensure that the project complies with the Building Regulations
- monitoring designers: taking reasonable steps to ensure that all designers are discharging their duties and promoting the statutory building safety objective
- golden thread: developing and/or contributing to the information that must be submitted at regulatory gateway points. Developing and maintaining a complete golden thread of information, and contributing to the handover process prior to occupation
- compliance: signing a declaration, along with the Principal Contractor, that the building complies with the Building Regulations.

The duties under the Building Safety Act will build on those required by the Construction (Design and Management) (CDM) Regulations 2015. The Principal Designer appointed under the CDM Regulations 2015 will be assumed to be the Principal Designer under the Building Safety Act.

The duties of the Principal Contractor will be

- leading construction: planning, monitoring and managing the construction phase and coordinating building safety to ensure that the project complies with the Building Regulations
- monitoring the supply chain: taking reasonable steps to ensure that contractors are meeting their duties and promoting the statutory building safety objective
- golden thread: developing and implementing the Construction Control Plan to ensure compliance with full plans and contributing to other information and record-keeping requirements
- compliance: obtaining sign-off from the Client and Principal Designer and/or regulator approval before deviating from approved plans during construction. Signing a declaration that the building complies with the Building Regulations and that an appropriate handover back to the Client has taken place.

As with the Principal Designer, the new duties of the Principal Contractor align with those allocated under the CDM Regulations 2015. The Principal Contractor appointed under the CDM Regulations 2015 will be assumed to be the Principal Contractor under the Building Safety Act.

The duties of designers, contractors and others involved with the project will include

- statutory objective: an over-arching duty to promote building safety and the safety of persons in and around the building
- mandatory occurrence reporting regime: complying with reporting requirements using the system established by the Client
- competence: only making and accepting appointments where the necessary skills, knowledge, experience and organisational capability are in evidence
- working together: cooperating, coordinating, sharing information and liaising with other duty holders and the regulator.

Similar duties apply to these roles under the CDM Regulations 2015. The Building Safety Act and the CDM Regulations 2015 are expected to work hand in hand so that all aspects of construction site safety and the future occupancy of the building are considered and hazards and risks eliminated or reduced to acceptable standards. Information is critical and must be handed over to those that require it.

The Building Safety Act includes provision for residents of existing higher-risk buildings to be included in the decision-making process and to be consulted where appropriate.

What are the key amendments proposed for the Building Act 1984 and the Building Regulations 2010?

It is proposed that the Building Safety Regulator establish and maintain a register of building inspectors (individuals) and building control approvers (either organisations or individuals). Individuals and organisations currently known as 'Approved Inspectors' who wish to continue undertaking building control work will need to register as 'Building Control Approvers'. The role of a registered building inspector being introduced in this Act is new. A registered building inspector is an individual who will be able to provide advice to local authorities or registered Building Control Approvers overseeing building work. Many inspectors in local authorities and Approved Inspectors are expected to transition to this role.

Other changes in the Act include

- removing the ability of a person carrying out building work on higher-risk buildings to choose their own building control body where building control is required
- changes to procedures for initial notices
- allowing the Secretary of State to make regulations as to how the new regulatory regime for higher-risk buildings will be applied (if at all) to a public body designated under sections 5 or 54 of the Building Act 1984
- providing information gathering powers on building control for the Regulator and Local Authority Building Control
- requiring the Building Safety Regulator to set up the new national electronic register/portal of information governing the work of registered Building Control Approvers
- amending the existing powers to give the Secretary of State the ability to designate bodies for the purposes of publishing criteria for and/or approving an insurance scheme
- applying a 15-year limitation period to section 38 of the Building Act 1984 (essentially, applicable where there is a breach of Building Regulations).

What will change regarding the competency of architects?

Architects will have to demonstrate relevant competency across all building disciplines. Disciplinary orders made against architects by the Professional Conduct Committee of the Architects Registration Board, the statutory regulator of architects in the UK, must be listed alongside an architect's entry in the Architects Register. This is to provide greater transparency to members of the public procuring architectural services as the Register entry, accessible through the Architects Registration Board's website, will show whether an architect has been disciplined recently. The Architects Registration Board will determine rules for the length of time that a disciplinary order will be listed on the Register, depending on the severity of the order.

Architects will also have to demonstrate continuing professional development in order to maintain their entry in the Architects Register.

What is the residents' panel?

Section 11 requires the Building Safety Regulator to establish a committee known as the residents' panel. The purpose of the panel is to ensure that residents have a voice in the work of the Building Safety Regulator, and the Building Safety Regulator has a broad power to consult the residents' panel on any of its functions that impact the residents of higher-risk buildings.

The panel must include residents, and may also have members from the following groups: owners of residential units (e.g. flats) in higher-risk buildings who are not occupying the property (who could, for example, be impacted by expenses relating to the higher-risk building); and groups or organisations that are representative of residents and/or non-occupying leaseholder owners, advocating for them or supporting them.

The Act also requires that the Building Safety Regulator must consult the residents' panel on certain matters that it is expected would be of particular interest and importance for residents of higher-risk buildings, specifically

- the residents' panel must be consulted on guidance to residents on their rights and obligations under the new regulatory regime for higher-risk buildings in occupation
- the residents' panel must be consulted on guidance relating to any duties relating to the provision of information, and so forth, to the regulator, residents and other persons where the guidance is about information provision to residents
- the residents' panel must be consulted on guidance relating to section 91 (residents' engagement strategy), section 92 (requests for further information), section 93 (complaints procedure operated by the Principal Accountable Person), section 95 (duties on residents and owners) or regulations made under any of those sections
- section 17 requires that the Building Safety Regulator consult the residents' panel on its strategic plan
- section 94 requires that the regulator consult the residents' panel on its system for dealing with complaints from residents that are escalated to the Building Safety Regulator.

What is the residents' engagement strategy?

Section 91 of the Building Safety Act 2022 requires the Principal Accountable Person to develop a residents' engagement strategy.

> The principal accountable person for an occupied higher-risk building must—
>
> (a) as soon as reasonably practicable after the relevant time prepare a strategy (a "residents' engagement strategy") for promoting the participation of relevant persons in the making of building safety decisions;
>
> (b) review the strategy at prescribed times, and revise it if they consider it necessary or appropriate to do so;

(c) in prescribed circumstances, consult relevant persons and prescribed persons on the strategy and take any representations made on the consultation into account when next reviewing the strategy;
(d) act in accordance with the strategy.

"relevant persons" are—

(a) residents of the higher-risk building who are aged 16 or over, and
(b) owners of residential units in the building.

Will officers, managers and employees of a corporate body be liable for prosecution should offences be committed?

Section 161 of the Building Safety Act 2022 addresses corporate responsibility as follows

Liability of officers of body corporate etc

(1) Where an offence under Part 2 or 4 committed by a body corporate is committed with the consent or connivance of any director, manager, secretary or other similar officer of the body corporate, or any person who was purporting to act in any such capacity, or is attributable to any neglect on the part of any such person, that person as well as the body corporate commits the offence and is liable to be proceeded against and punished accordingly.
(2) In subsection (1) "director", in relation to a body corporate whose affairs are managed by its members, means a member of the body corporate.
(3) Subsection (1) also applies in relation to a body that is not a body corporate, with the substitution for the reference to a director of the body of a reference in the case of a partnership, to a partner; in the case of an unincorporated body other than a partnership, where the body's affairs are managed by its members, to a member of the body; in any other case, to a member of the governing body.
(4) Where:

    (a) a relevant company is an accountable person for a higher-risk building (within the meaning of Part 4), and
    (b) one or more (but not all) directors of the relevant company have been appointed for a building safety purpose and are entitled to remuneration from the company,

    this section, so far as relating to Part 4, does not apply in relation to a director who is not entitled to remuneration from the relevant company.

What are the fines and sentencing provision under the Building Safety Act for failing to comply with regulatory duties?

Both the body corporate and individual officers of the company are liable for the offences committed under the Building Safety Act.

Those offences include behaviour that constitutes intentional obstruction of someone who is 'exercising a relevant building function' (section 23) or where false or misleading information is provided to the Building Safety Regulator (section 24).

There is a host of responsibilities in respect of higher-risk buildings (HRBs), set out in Part 4 of the Act, that could lead to personal liability under section 161, including

- allowing occupation of a HRB without the relevant completion certificate (section 76)
- allowing occupation of a HRB without registration (section 77)
- failure to action or comply with a Compliance Notice (section 99)
- failure to apply for a Building Assessment Certificate when directed to do so (section 79)
- failure to display various notices and certificates in the building (section 82)
- failure to operate a mandatory reporting system (where required) related to the safety of an occupied building (section 87)
- where there is a change in Accountable Persons and the previous Accountable Person fails to provide prescribed information and documents to the new Accountable Person (section 90)
- where an Accountable Person breaches a relevant requirement that places people in or around the building at critical risk (section 101).

The Building Safety Regulator has the power to issue a Compliance Notice requiring remedial steps to be undertaken by the Accountable Person within a specified timeframe where the regulator considers the Accountable Person has breached one of its obligations. Failure to comply with the notice within the specified time period is a criminal offence.

Fines for breaches can be unlimited and many offences carry custodial sentences of up to 1 year if the case is heard in the Magistrates' Court and 2 years if the case is heard in the Crown Court.

The Building Safety Act 2022 introduces Building Liability Orders and Information Orders. What are these?

Developers often create subsidiary companies, such as 'shell' companies or special purpose vehicles (SPVs), for the sole purpose of carrying out particular development projects. Once the project has concluded, it is generally commonplace for developers to transfer any assets out of the subsidiary company to other parts of its business and then look to dissolve the company.

Such a corporate structure is useful for developers but it often leads to a situation where the company that undertook the works on a project is either dissolved or holds limited assets of value. This can significantly impact a Claimant's recovery position if defects are found post-completion of the project.

The introduction of Building Liability Orders and Information Orders aims to address the above situation by broadening the Claimant's recovery rights and introducing a way of extending liability beyond the original subsidiary company to its associated companies.

Under section 130 of the Building Safety Act 2022, the High Court has the right to make a Building Liability Order. Once ordered, a Building Liability Order extends the specific liabilities of one body corporate to an associated company/companies, making them jointly and severally liable.

A Building Liability Order may be made in relation to any 'relevant liability' (s.130(3)), which is defined as a liability that is incurred under the Defective Premises Act 1972 (DPA) or resulting from a 'building safety risk' – which is defined as a 'risk to the safety of people in or about the building arising from the spread of fire or structural failure' (s130(6)).

Applicants will need to show that there is a 'just and equitable' reason for the High Court to issue a Building Liability Order.

In addition to Building Liability Orders, section 132 of the Building Safety Act 2022 introduces the right for a Claimant to apply for an Information Order. If granted, this order requires a specific company to give specified information or documents related to its associated companies (s.132(2)). This may be a powerful tool for potential Claimants to use in order to determine whether there is merit in applying for a Building Liability Order and/or which company/companies a Building Liability Order application should be targeted at. To obtain an Information Order, the applicant will need to show that it requires the specified information or documents in order to enable it to consider whether to apply for a Building Liability Order.

Are there any costs involved when the Building Safety Regulator acts under the Building Safety Act 2022?

Yes, there can be.

The Building Safety (Regulator's Charges) Regulations 2023 enables the Building Safety Regulator to charge for certain functions under the Building Act 1984 (as amended) and the Building Safety Act 2022.

In general terms, applications for building control approval are chargeable, with both an application fee and hourly charges of approximately £151/h (as of April 2025) for Regulator staff.

In respect of higher-risk buildings, the Regulator can charge for

- applications and registrations of higher-risk buildings
- inspections
- reviews and appeals.

Again, costs will be fixed fees for applications – for example, completion certificates, higher-risk building registrations – and hourly fees for any further work needed by the Regulator.

If a person/company is thought to be contravening the legislation and the Regulator decides to investigate and take formal action to ensure compliance, there will be a charge for such intervention. This is similar to the existing 'fee for fault' charged by the Health and Safety Executive for health and safety investigations.

Reviews and appeals are also chargeable.

Hourly fees have been set at £151 ((as of April 2025). All fees levied must be paid within 30 days of receipt of invoice.

Works required purely for a disabled person will not be subject to any fees.

REFERENCES

Building Safety Act 2022.
Information and guidance documents relating to the Building Safety Act 2022: Department for Levelling Up, Housing and Communities.
Information on the Building Safety Regulator: Health and Safety Executive.
The Building Safety (Regulator's Charges) Regulations 2023.
The Higher-Risk Buildings (Descriptions and Supplementary Provisions) Regulations 2023.
The Higher-Risk Buildings (Management of Safety Risks etc) (England) Regulations 2023.

TOP TIPS

- Survey your properties to establish whether you have higher-risk buildings. Specific legal requirements apply to these buildings.
- Review all fire safety procedures and ensure that information is made available to all who will need it and that it is shared among other duty holders.
- Update Fire Risk Assessments.
- Appoint only competent people as duty holders.
- Know the definition of high-rise and higher-risk buildings and whether the Building Safety Act applies.
- Appoint an Accountable Person and, where necessary, a Principal Accountable Person. Put the appointment in writing.
- Ensure that an Accountable Person is always in post.
- Make appointments in writing and specify responsibilities.
- Establish ways of introducing the 'golden thread' of information for all information relevant to the safety of the building.
- Communicate with residents if you own or manage residential buildings especially about fire safety matters and share information.
- Communicate with all building occupiers and others who are responsible for parts of the building and share information.
- Be proactive and take control of your duties under the new legislation. Seek competent advice where necessary.
- Review the construction materials specified and ensure that they are fit for purpose and meet specifications.

- Check building and refurbishment works for compliance with the tender or contract documents.
- Do not turn a blind eye to managing the building works or the building once occupied. Legal duties cannot be delegated.
- Manage refurbishment works closely and appoint a competent design and project management team.

The Building Safety Act 2022: Questions and Answers: A practical approach to design and management of higher-risk buildings

Pat Perry

ISBN 978-1-83662-467-7
https://doi.org/10.1108/978-1-83662-464-620251004

Chapter 4
Enforcement sanctions

The Building Safety Act 2022 allows for the Building Control Authority to issue a 'stop notice'. What is this?

A stop notice is an enforcement mechanism that can be used by the Building Control Authority to control serious contraventions of the Building Regulations where the Client or Contractor has failed to respond to a compliance notice.

The Building safety Act 2022, section 35C states:

(1) The Building Control Authority may give a stop notice to a person appearing to the authority to be in control of any work if it appears to the authority that—

  (a) the carrying out of the work would contravene a provision of building regulations prescribed for the purposes of this paragraph;
  (b) a compliance notice relating to the work has been contravened; or
  (c) the work contravenes a provision of building regulations or a requirement imposed by virtue of such a provision, and the risk of serious harm condition is met.

(2) For the purposes of subsection (1)(c) the "risk of serious harm condition" is that use of the building in question without the contravention having been remedied would be likely to present a risk of serious harm to people in or about the building.

(3) A "stop notice" is a notice prohibiting, either immediately or from a specified time, the carrying out of specified work until the occurrence of such of the following as may be specified—

  (a) the taking of specified steps;
  (b) the occurrence of specified circumstances;
  (c) the remedying of a specified contravention or the matters giving rise to it.

(4) Where a stop notice is contravened, the person to whom the notice was given commits an offence.

(5) A person guilty of an offence under this section is liable—

  (a) on summary conviction, to imprisonment for a term not exceeding the maximum summary term for either-way offences or a fine (or both);
  (b) on conviction on indictment, to imprisonment for a term not exceeding two years or a fine (or both);

  and (in either case) is liable on summary conviction to a further fine not exceeding level 1 on the standard scale for each day on which the default continues after the initial conviction.

(6) It is a defence for a person charged with an offence under this section to prove that the person took all reasonable precautions and exercised all due diligence to avoid the contravention of the stop notice.

When could a stop notice be served?

Stop notices can only be issued in certain situations. These are

- that carrying out work would contravene certain prescribed Building Regulations
- an existing compliance notice has not been complied with
- works carried out contravene regulations and could potentially cause a risk of serious harm (no 12-month time limit attached to this notice).

The stop notice will require work to which the stop notice relates to be stopped altogether until remedial action has been taken. A stop notice can apply to a whole site, a part of a site or a specific area of work. These powers are available for a breach that occurred after 1 October 2023.

What is a 'compliance notice'?

A compliance notice is issued either by the Building Safety Regulator in respect of higher-risk buildings or the Building Control Authority in relation to all other building works, when it has been determined that the requirements of any of the Building Regulations have not been met.

The Building Regulations etc. (Amendment) (England) Regulations 2023 amend Regulation 47 of the Building Regulations 2010 and section 35 Building Act 1984 to state

47B. (1) A compliance notice given by a building control authority must be in writing and state—

(a) that it is a compliance notice under section 35B of the Act;
(b) the date the notice is given;
(c) the name or description of the recipient; and
(d) the consequences of failing to comply with the notice.

(2) A compliance notice must additionally—

(a) describe the issue to which the notice relates;
(b) identify the provision of building regulations, or the requirement imposed by virtue of building regulations, to which the compliance notice relates;
(c) provide details of the nature of the contravention or likely contravention;
(d) state that the notice may be withdrawn by the building control authority;
(e) state that the recipient may apply to the First-tier Tribunal under section 39A (3) of the Act for an extension of the specified period for doing anything set out in the notice and state that any application must be made within 21 relevant days of the date the notice is given;
(f) state that the recipient may appeal to the First-tier Tribunal under section 39A (1) of the Act and state that any appeal must be made within 21 relevant days of the date the notice is given.

(3) A compliance notice must relate to no more than one contravention or likely contravention of building regulations or a requirement imposed by virtue of building regulations.

If there have been multiple contraventions or likely contraventions, then multiple compliance notices or stop notices will need to be issued. If multiple notices have been issued, then each notice will apply to a specific contravention relating to the relevant Building Regulation.

Each contravention will require bespoke remedies and compliance periods. This approach of requiring an individual notice for each contravention or likely contravention provides clarity about what is required from the developer/owner/builder and allows flexibility in managing the construction of complex sites in particular.

Who would receive the enforcement notices?

Where a Building Control Authority gives a compliance notice or stop notice, the authority must take reasonable steps to notify the following persons as to the giving of the notice

- the Client, the Principal Contractor (or sole contractor) and the Principal Designer (or sole or lead designer) for the work to which the notice relates
- where the Building Safety Regulator is the Building Control Authority, the local authority for the area in which the building is situated or the proposed building is to be situated
- where the notice relates to a contravention of Part B of Schedule 1 (Fire Safety) in relation to a building to which the Regulatory Reform (Fire Safety) Order 2005 applies or will apply after completion of the work, the enforcing authority (within the meaning of article 25 of that Order) for the building or the proposed building.

Can a compliance notice be appealed?

Yes. If you disagree with the reason for being issued with a compliance notice you can appeal to the First-tier Tribunal. The notice of the appeal must be submitted within 21 days of receipt of the compliance notice.

You can also appeal the notice if you feel that it has given insufficient time for the works to be carried out – that is, you can request an extension of time. Again, this has to be submitted within 21 days of receipt of the original notice.

The First-tier Tribunal may allow an appeal if it is satisfied that the giving of the compliance notice or stop notice in question was wrong on one or more of the following grounds

- the notice was given based on an error of fact
- the giving of the notice was wrong in law
- the giving of the notice was unreasonable
- the notice was given without following the procedures set out in the Act or regulations made under that Act (except Regulation 47D(3)).

Is it an offence to ignore a compliance notice?

Yes. A person who, without reasonable excuse, contravenes a compliance notice commits an offence.

Section 35B Building Act 1984 (added by the Building Safety Act 2022, section 38) states

> (5) A person guilty of an offence under this section is liable—
>
> (a) on summary conviction, to imprisonment for a term not exceeding the maximum summary term for either-way offences or a fine (or both);
> (b) on conviction on indictment, to imprisonment for a term not exceeding two years or a fine (or both)
>
> and (in either case) is liable on summary conviction to a further fine not exceeding level 1 on the standard scale for each day on which the default continues after the initial conviction.

What is a 'section 36' notice under the Building Act 1984?

The local authority, as Building Control Authority, may (under section 36 of the Act) serve an enforcement notice on the owner of a building requiring them to alter or remove work that contravenes the Building Regulations.

If the owner does not comply with the notice, the local authority has the power to undertake the work itself and recover the costs from the owner.

A section 36 notice cannot be served more than 12 months after the date of completion of the building work.

The Building Safety Regulator, as the Building Control Authority for higher-risk buildings, has similar powers to serve notices for contraventions.

Can either a stop notice or a compliance notice be withdrawn by the Building Control Authority?

Yes. Compliance and stop notices can be withdrawn if either they have been served incorrectly, are unnecessary or if the works have been completed. Section 47 Building Regulations 2010 states

> (1) A building control authority may withdraw a compliance notice or stop notice.
> (2) Where a building control authority withdraws a compliance notice or stop notice it must as soon as reasonably practicable give the recipient a notice in writing to that effect.

Time limits for enforcement action have changed significantly under the amendments brought in by the Building Safety Act 2022 and its secondary legislation. What are the differences?

Time limits for enforcement were increased by the Building Safety Act. Prior to the Building Safety Act, the relevant period to take an enforcement action (e.g. giving notice requiring removal or alteration of non-compliant work) was 12 months.

This has increased significantly to 10 years from the date of completion of the works in question as per section 36(4) of the amended Building Act 1984.

In addition, if there is a contravention of Building Regulations (e.g. Building Act 1984, s.35), there is no longer a time limit for prosecution.

Can individuals be personally liable for contraventions of the Building Safety Act 2022?

Yes, individuals complicit in breaches may now be found liable for such contraventions in addition to a body corporate.

Under s.112A of the Building Act 1984 (as amended by s.40 of the Building Safety Act), where an offence is committed by a body corporate, individuals within that organisation (e.g. directors, managers, secretary and similar officers) may also be prosecuted for an offence where the breach was committed with their consent or connivance or as a result of their neglect.

Enforcement authorities will consider management arrangements and the role played by individual directors and managers in relation to an offence as they can be held responsible for their actions or inactions.

What are Remediation Orders?

Remediation Orders were introduced under the Building Safety Act 2022 to ensure that essential remediation work required to remedy relevant defects – that is, those relating to fire and structural safety – takes place without delay. Regulatory bodies and leaseholders can apply to the First-tier Tribunal for a Remediation Order to compel a relevant landlord to remedy relevant defects by a specified time.

Section 123 of the Building Safety Act 2022 provides 'interested persons' with a right to apply for a Remediation Order requiring a relevant landlord to remedy specified relevant defects in a specified relevant building by a specified time.

'Interested persons' include

- the Building Safety Regulator
- a local authority
- a fire and rescue authority
- a person with a proprietary interest in a relevant building (such as a leaseholder)
- the Homes and Communities Agency.

The property needs to be a 'relevant building', which means a building or any self-contained part of a building in England

- containing at least two dwellings
- with a height of 11 m, or five storeys or more.

Certain buildings are excluded, such as leaseholder-owned buildings or those on commonhold land. The definition of a 'relevant building' for the purposes of Remediation Orders is not the same as the definition of 'higher-risk buildings' set out in the Building Safety Act.

A Remediation Order is made against the 'relevant landlord' who is a landlord under a lease of a relevant building who is required to repair or maintain anything relating to the relevant defect. This includes management companies, freeholders and superior landlords if they have repairing/maintenance obligations.

A Remediation Order needs to remedy a 'relevant defect'. This is a defect arising as a result of anything done (or not done) in connection with the relevant works and causes a building safety risk.

A 'building safety risk' includes risks to the safety of people in or about the building arising from the spread of fire, or the collapse of the building or any part thereof.

Remediation Orders are issued in particular to deal with flammable external cladding and fire safety risks – for example, lack of fireproof doors to common areas, lack of working fire alarm systems, and so forth.

Remediation legal case orders freeholders to pay

In a landmark legal ruling in 2024, the government secured a court order compelling a multibillion-pound pension fund's property business to finally remediate dangerous fire safety defects in a residential tower block. The judgment marks a major victory for leaseholders caught in holdups over who should pay for essential cladding and safety repairs.

Stevenage high-rise tower block – a higher-risk building

The case centres on Vista Tower, a high-rise apartment building in Stevenage owned by freeholder Grey GR. Despite fire hazards first being identified in 2019, Grey GR – a subsidiary ultimately controlled by Railpen's £34 billion pension scheme for railway workers – had failed to initiate comprehensive remediation works to make the building safe.

As the impasse dragged on, with leaseholders facing disruption, uncertainty and diminished property values, the government's Department for Levelling Up, Housing and Communities (DLUHC) took the exceptional step of launching legal action under new Building Safety Act powers.

Court acts in residents' favour

Following a trial in March 2024, the firm owned by Railpen was ordered to pay for cladding remediation at the 16-storey building by the First-tier Tribunal Property Chamber.

Remediation costs amounted to around £15 million. This includes a reimbursement of £3.7 million to the government and Homes England that was used to cover fire safety fixes in February 2024.

The Court issued a legally binding 'remediation order' compelling Grey GR to resolve the fire safety deficiencies at Vista Tower according to a strict, court-mandated timeline and work plan.

'This decision is a victory for leaseholders...who have lived with uncertainty for far too long,' said Levelling Up Secretary, Michael Gove. 'It should serve as a warning to all building owners – if you fail to fix your unsafe buildings, we will see you in court.'

What is a Building Liability Order?

The Building Safety Act 2022 provides the High Court with the power to issue a new remedy, through which the costs of rectifying building safety risks can be obtained from companies that were not directly involved with the original construction. Once ordered, subsidiaries, shell companies or special purpose vehicles (which are often dissolved after the works are complete) as a means to reduce exposure to risk, may be jointly and severally liable for remedial costs.

Building Liability Orders may be made in relation to a 'relevant liability'. A relevant liability means a liability incurred

- under the Defective Premises Act 1972 (DPA) where either workmanship or materials result in a dwelling being unfit for habitation
- under section 38 of the Building Act 1984 where there has been a breach of Building Regulations that has caused damage
- as a result of a building safety risk – that is, a risk to the safety of people in or about a building arising from the spread of fire or structural failure.

The court will only order a Building Liability Order against a 'specified' body corporate. A body corporate will be specified only if it is, or was, associated with the original party at any time from the date of the beginning of the works up to the date the Building Liability Order is made.

A body corporate is associated with another body corporate where one of them is controlled by the other (e.g. holds at least half of the issued share capital or holds at least half the voting rights), or where a third party controls them both.

How has enforcement for non-compliance with Building Regulations changed under the Building Safety Act 2022?

On 1 October 2023, the Building Safety Act significantly changed and extended the rights of enforcement and potential sanctions for non-compliance with Building Regulations. The two key elements to these changes are as follows.

Increased enforcement period

Where work is not in compliance with the Building Regulations, then the local authority or the Building Control Authority can require the works to be removed or altered. Prior to 1 October 2023, such a requirement could be made up to 12 months after the works had been completed. This enforcement period has been extended to 10 years after the works have been completed (section 39(3) of the Building Safety Act 2022).

Increased fines and penalties

Prior to the enactment of the Building Safety Act 2022, non-compliance with Building Regulations was a summary-only offence with a fine up to £5000 plus a maximum fine of £50 per day of default after conviction.

Under the new Act, the offence of non-compliance now carries an unlimited fine and/or imprisonment of up to 2 years plus a further fine per day of default after conviction of up to £200 per day (section 39 of the Building Safety Act 2022).

Non-compliance covers work where

- plans or an application were required and not deposited or made and approved
- work is carried out notwithstanding any rejection of the deposited plans or application made and approved
- work is otherwise carried out in a way that is contrary to the plans deposited or application made and approved.

Remember: the new enforcement provisions apply to all building works and not just those that involve higher-risk buildings.

Offence	Scope of power	Penalty	Reference
Providing false or misleading information to the Building Safety Regulator	Criminal offence, Triable as an either-way offence	Unlimited fine, 1 year imprisonment (magistrates), 2 years imprisonment (crown)	Section 24 BSA22
Contravention of Building Regulations	Criminal offence, Triable as an either-way offence	No time limit, Unlimited fine, Fine for each day the contravention continues (£200)	Section 36 BA84
Failure to comply with compliance and stop notice	Criminal offence	Unlimited fine, 1 year imprisonment (magistrates), 2 years imprisonment (crown)	Section 35B & C BA84
Removal of offending work	Time limit extended to 10 years	Notice to remove contravening work, Chargeable work	Section 36 BA84

Directors and managers can also be found liable for any of the criminal offences mentioned.

Compliance and stop notices can be served without the need to go to a magistrate's court in the first instance.

A breach of the Building Regulations removes altogether the time limit for prosecutions under section 35 of the Building Act 1984, and will have an increased penalty, with a maximum penalty of an unlimited fine and up to 2 years in prison.

Compliance notices can be served against any breach of the Building Regulations and require corrective action by a set date. Failure to comply will be a criminal offence.

Stop notices can be served against a breach of specified Building Regulations, such as a breach of a compliance notice and any breach of Building Regulations that is likely to cause 'serious harm'. A stop notice requires that all specified work is stopped until the contravention is rectified and failure to comply will be a criminal offence.

Section 36 of Building Act 1984 (notice for remedial action) extends the time limit for enforcement from 2 years to 10 years.

Building Act 1984 Compliance Notice (Example)

To: [Owner/Occupier Name]

Address: [Property Address]

Date: [Date of Notice]

Details of Non-Compliance:

Description of Work: [Detailed description of the building work that is non-compliant with Building Regulations]

Breached Regulation: [Specific Building Regulation number that has been violated – one contravention per Notice]

Reason for Non-Compliance: [Explanation of how the work fails to meet the Building Regulations]

Action Required:

Remedy: [Specific actions required to bring the building work into compliance, including any necessary alterations, modifications, or removal of non-compliant elements]

Completion Date: [Date by which the corrective work must be completed]

Further Actions:

Inspection Request:

You must contact the Building Control Authority to arrange an inspection of the completed remedial work.

Failure to Comply:

If you fail to comply with this notice by the specified date, further enforcement action may be taken, including potential legal proceedings.

Contact Details:

Building Control Authority: [Name of Local Authority Building Control Department]

Contact Person: [Name of Contact Officer]

Phone Number: [Contact Phone Number]

Email Address: [Contact Email Address]

Signed:

Position: **Date:**

Additional recipients (those indicated have received a copy of this Notice)

Client:

Building Owner:

Principal Designer:

Principal Contractor:

Local Authority:

Fire Authority:

The Building Safety Act 2022: Questions and Answers: A practical approach to design and management of higher-risk buildings

Pat Perry
ISBN 978-1-83662-467-7
https://doi.org/10.1108/978-1-83662-464-620251005

Chapter 5
Competency standards

Why is there a new emphasis on competency standards in the construction industry?

Dame Judith Hackitt's Independent Review of Building Regulations and Fire Safety, following the Grenfell Tower fire in 2017, identified that levels of competency in the construction industry were woefully low and that compliance with Building Regulations and other legislation was often considered unimportant and this, in turn, compromised building safety.

She found the key issues underpinning the system failure included the following.

- Ignorance – regulations and guidance are not always read by those who need to, and when they do the guidance is misunderstood and misinterpreted.
- Indifference – the primary motivation is to do things as quickly and cheaply as possible rather than to deliver quality homes that are safe for people to live in. When concerns are raised, by others involved in building work or by residents, they are often ignored. Some of those undertaking building work fail to prioritise safety, using the ambiguity of regulations and guidance to game the system.
- Lack of clarity on roles and responsibilities – there is ambiguity over where responsibility lies, exacerbated by a level of fragmentation within the industry, and precluding robust ownership of accountability.

Dame Hackitt found that the competency of those involved in higher-risk buildings was particularly poor and that any understanding of building safety was by and large inadequate.

One of her key recommendations in her final report: 'Building a Safer Future' – chapter on 'Competence' – was

> The construction sector and fire safety sector should:
>
> a. demonstrate more effective leadership in relation to developing a responsible approach to delivering building safety and integrity;
> b. work with other sectors to learn and translate good practice and implement it within the sector; and
> c. develop continuous improvement approaches to competence levels.

In his final report of the Grenfell Tower Public Inquiry, Sir Martin Moore-Bick, stated

> As architect Studio E was responsible for the design of the external wall and for the choice of the materials used in its construction. Although the TMO [Tenant Management Organisation] as the client wanted to reduce the cost by using ACM rainscreen panels, it was the responsibility of Studio E to determine whether the use of such material would enable the building to comply with functional requirement B4(1) of the Building Regulations and advise the TMO accordingly. Its failure to recognise that ACM was dangerous and to warn the TMO against its use represented a failure to act in accordance with the standard of a reasonably competent architect. It also failed to recognise that Celotex insulation was combustible and not suitable for use on a building over 18 metres in height in accordance with the statutory guidance. Studio E therefore bears a very significant degree of responsibility for the disaster.
>
> We have identified many other respects in which Studio E failed to meet the standards of a reasonably competent architect, of which the following are the most significant.
>
> - It failed to ensure that Exova completed the fire safety strategy for the refurbished building or advise Rydon and the TMO that it should be required to do so.
> - It failed to understand that it was responsible for design work carried out by sub-contractors and so did not check Harley's designs to ensure that on completion the building would comply with the Building Regulations.
> - It did not devise a proper cavity barrier strategy or check Harley's designs for the cavity barriers.
> - It failed to produce detailed drawings of the window reveals or to notice that the materials specified for the window infill panels were unsuitable.

Other members of the design and professional team working on the Grenfell Tower refurbishment project were singled out for incompetency and, again, Sir Moore-Bick stated

> Exova also bears considerable responsibility for the fact that Grenfell Tower was in a dangerous condition on completion of the refurbishment. Our most serious criticism is that it failed to produce a final version of the fire safety strategy for the refurbished building and that it failed either to draw that fact to the attention of the design team or to warn it about the potential consequences. None of those responsible for drafting the fire safety strategy visited Grenfell Tower; the only site visit by a member of Exova's staff took place at a preliminary stage. Exova's attitude was wholly inconsistent with the careful approach to matters affecting the safety of life to be expected of a reasonably competent fire engineer.
>
> We consider that the principal contractor, Rydon, also bears considerable responsibility for the fire. It gave inadequate thought to fire safety, to which it displayed a casual attitude throughout the project and its systems for managing the design work did not ensure that its sub-contractors and consultants properly understood their different responsibilities. Rydon itself did not understand where responsibility for individual decisions lay and as a result it failed to co-ordinate the design work properly.

> Rydon had an inexperienced team on the refurbishment that did not have sufficient knowledge of the Building Regulations or Approved Document B. It relied entirely on its cladding sub-contractor, Harley, to draw its attention to any errors in the design, but it did not specifically ask Harley to assess Studio E's work. It failed to take proper steps to investigate Harley's competence and ensure that it was competent to undertake the work and capable of providing the services required of it. It was complacent about the need for fire engineering advice and took the decision not to retain Exova without consulting the TMO, Studio E or Artelia. Its understanding of the work already carried out by Exova was superficial; as a result, it failed to realise that the fire safety strategy had not been completed.
>
> Harley itself failed in many respects to meet the standards to be expected of a reasonably competent cladding contractor and it too bears a significant degree of responsibility for the fire. It did not concern itself sufficiently with fire safety at any stage of the refurbishment and appears to have thought that there was no need for it to do so, because others involved in the project, and ultimately building control, would ensure that the design was safe. It failed to ask the kind of questions about the materials being considered that a reasonably competent cladding contractor would have asked. It was induced to buy Reynobond 55 PE panels partly by its existing relationship with Arconic and the cladding fabricator, CEP Architectural Facades, with which it was able to negotiate a favourable price. Its staff were unaware of the requirements of the Building Regulations relating to fire safety, the guidance in Approved Document B or industry guidance and did not understand the underlying testing regime.

Lack of competency was also identified with fire risk assessors tasked with carrying out fire risk assessments in higher-risk buildings.

Issues of competency played a large part in the Grenfell Tower fire tragedy and recommendations were made to enhance and improve competency standards across a range of professional disciplines.

Who is responsible for checking the competency standards of those working on a construction project?

Everybody who commissions building works has a duty to check competency standards.

The new section 11F to the Building Regulations 2010 states

> (1) Any person carrying out any building work or any design work must have—
>
> (a) where the person is an individual, the skills, knowledge, experience and behaviours necessary,
>
> (b) where the person is not an individual, the organisational capability,
>
> to carry out—
>
> (i) the building work in accordance with all relevant requirements;
>
> (ii) the design work so that the building work to which the design relates, if built, would be in accordance with all relevant requirements.

(2) Any person carrying out any building work as a contractor or any design work as a designer must have—

(a) where the person is an individual, the skills, knowledge, experience and behaviours necessary,

(b) where the person is not an individual, the organisational capability, to fulfil the duties of a contractor or designer, as the case may be, under these Regulations in relation to the work.

(3) The requirements in paragraphs (1) and (2) do not apply to an individual (T) who is in training to fulfil those requirements.

(4) The person who asked T to carry out any building work or, as the case may be, any design work must ensure T is adequately supervised when carrying out the work.

(5) A person who is in training to fulfil the requirements of a principal contractor or a principal designer may not be appointed as a principal contractor or a principal designer.

Who checks the competency of the Principal Designer and the Principal Contractor, as required by the Building Regulations?

A Client will need to appoint a Principal Designer and Principal Contractor to any building project that is likely to have more than one contractor.

The Client is therefore responsible for being satisfied about the competency of the persons so appointed.

Both the Principal Designer and Principal Contractor are responsible for demonstrating to the Client that they have the competencies to carry out the roles.

What are the key competency requirements?

The key competency requirements are

- skills
- knowledge
- experience
- behaviours.

Are there any recognised published standards against which to check competency of organisations and individuals?

Yes. Many professional bodies such as the Royal Institute of British Architects, the Construction Leadership Council and the Fire Protection Association have been working towards issuing competency standards for their own professions.

The British Standards Institute have issued Publicly Available Specifications for Competency, namely

- PAS 8671: 2022 Built environment – Framework for competence of individual Principal Designers – Specification

- PAS 8672: 2022 Built environment – Framework for competence of individual Principal Contractors – Specification
- PAS 8673: 2022 Built environment – Competence requirements for the management of safety in residential buildings – Specification.

In addition to the above, the BSI have issued a new British Standard

BS 8670–1:2024 Competence frameworks for building safety – Core criteria: Code of practice.

What are the steps I need to take to check competency levels of those undertaking the legal duties of Principal Designer?

When a Principal Designer is appointed under the Building Regulations, they are responsible for planning, managing and monitoring design work, and cooperating, coordinating and communicating with others to ensure the design work, if built, complies with the Regulations.

So, the Principal Designer must be able to demonstrate good management and communication skills.

Core competency criteria include

1. Behavioural competence for building safety
    - act ethically and contribute to safe outcomes
    - demonstrate effective leadership, teamwork and communication
    - manage individual and contribute to organisational competence
    - demonstrate personal responsibility and accountability
    - understand and comply with duty of care to others
2. Legislative and regulatory framework for compliance

    The specific competencies that the PAS 8671 lists, which are needed to demonstrate this, relate to understanding
    - the duties and behaviours required, the purpose, structure and scope of the legislative and regulatory framework
    - how to find and apply the relevant information in the legislative and regulatory framework.
3. Management of design work compliance

    Should be able to manage, monitor and coordinate designers and design work in relation to compliance. The specific competencies are
    - the ability to demonstrate understanding of a project and its requirements, specifically how this can impact or affect design work compliance, and the effect on designers' duties and contracted responsibilities in light of the details of the project

- the ability to strategise effectively, analyse, manage and monitor gaps in designers' competence and identify risks to compliance, and control changes affecting design work compliance
- the ability to manage the recording and maintenance of evidence of design work compliance.

The PAS 8671 describes 'management' as

- setting the plan for achieving design work compliance
- controlling changes to the plan and monitor risks to design work compliance
- coordinating designers' work related to achieving the plan
- liaising with principal contractors about design work compliance
- monitoring progress against the plan through to the end of the design phase

4. Technical framework for compliance

   Should understand general principles of building design, construction and the "core criteria for building safety" set out in the BSI Flex 8670 so they can assess, manage and appraise evidence of design work compliance in order to build design team consensus on compliance.

Is the Principal Designer expected to be an expert in all design disciplines?

No. A Principal Designer is not expected to be an expert in every design specialism, but they are expected to know enough about the Building Regulations to assess whether a building design will comply with all relevant regulations.

Specific competencies relating to this comprise being able to understand the purpose, scope and structure of the technical framework, how to find and apply the relevant information contained in it, and how to use information management systems effectively to maintain design work compliance.

Is checking competency standards an ongoing event during the project?

It would be sensible to include procedures in the overall project management plan for the competency of duty holders to be regularly reviewed, especially where complex designs are being created and competency expectations may change.

The Client will need to be reassured that the duty holders appointed remain competent to perform their roles and while this may not require a full re-assessment a continuous review of skills, knowledge, experience and behaviours needs to be built in to the culture of the project.

Can an organisation take on the role of Principal Designer?

Yes. An organisation can fulfil the duties allocated to the Principal Designer role but must nominate a 'designated individual' to manage its functions. The designated individual should possess the competencies outlined in PAS 8671.

Organisations must ensure their workforce has the skills, knowledge, experience and behaviours for their roles. PAS 8671 can help businesses assess the competence of the individuals working for them.

Organisations must have the necessary 'organisational capability'. This requires them to actively monitor and supervise their people, while also providing sufficient time and resources to do the job.

Do Principal Designers need to demonstrate any additional competencies if they are appointed to a project involving a higher-risk building?

PAS 8671 states that there are additional competencies required for working on higher-risk buildings (HRBs), as follows

> Principal Designers shall be able to:
>
> (a) analyse their understanding of the legislative and regulatory framework related to designing HRBs, including in relation to prescribed procedures and information; and
> (b) evaluate duties of other duty holders in the context of HRB projects for the ways they could affect the Principal Designer's duties and design work compliance; and
> (c) evaluate the system of regulated procedures and information related to working on HRBs and their implications for the performance of the Principal Designer's duties, including contributing to the golden thread of information, reporting safety occurrences to the Building Safety Regulator, and making competence and compliance declarations; and
> (d) evaluate ways to establish and maintain throughout the construction phase a system for inspecting HRB design work for safety occurrences and promptly reporting safety occurrences; and
> (e) create ways to instruct reporting persons about the system for mandatory safety occurrence reporting related to working on HRBs; and
> (f) evaluate building safety risks in connection with the regulated system of mandatory safety occurrence reporting sufficient to report events related to structural safety, fire safety and other prescribed matters in HRBs; and
> (g) evaluate the potential during inspections of design work throughout the construction phase for new building safety risks to emerge and for the assessed severity of potential impacts of all risks to change, including to become safety occurrences; and
> (h) evaluate design work compliance sufficient to follow prescribed procedures for declaring compliance; and
> (i) understand information management systems sufficient to:
>
> - contribute to the collation, accurate maintenance, and accessibility of digital records in the golden thread of information; and
> - arrange the establishment and maintenance of the mandatory occurrence reporting system.

PAS 8673: 2022 Built environment – Competence requirements for the management of safety in residential buildings – sets out the additional competencies needed for those working on and managing residential projects.

The Building Safety Act 2022 introduces obligations for Accountable Persons to be responsible for assessing and managing building safety risks (i.e. the spread of fire and structural failure). To meet the duties set out in the Act, some Accountable Persons might find it necessary to improve their arrangements, including their competence or of those that support them, to assess and manage building safety risks.

Does the Client have to carry out additional assessments where the Principal Designer is also carrying out the Principal Designer role under the Construction (Design and Management) (CDM) Regulations 2015?

Yes, because the competency requirements under the CDM Regulations 2015 are different from those required under the Building Safety Act 2022.

Regulation 8 CDM states

> 8.—(1) A designer (including a principal designer) or contractor (including a principal contractor) appointed to work on a project must have the skills, knowledge and experience, and, if they are an organisation, the organisational capability, necessary to fulfil the role that they are appointed to undertake, in a manner that secures the health and safety of any person affected by the project.
>
> (2) A designer or contractor must not accept an appointment to a project unless they fulfil the conditions in paragraph (1).
>
> (3) A person who is responsible for appointing a designer or contractor to carry out work on a project must take reasonable steps to satisfy themselves that the designer or contractor fulfils the conditions in paragraph (1).
>
> (4) A person with a duty or function under these Regulations must cooperate with any other person working on or in relation to a project, at the same or an adjoining construction site, to the extent necessary to enable any person with a duty or function to fulfil that duty or function.
>
> (5) A person working on a project under the control of another must report to that person anything they are aware of in relation to the project which is likely to endanger their own health or safety or that of others.
>
> (6) Any person who is required by these Regulations to provide information or instruction must ensure the information or instruction is comprehensible and provided as soon as is practicable.

The CDM Regulations 2015 are primarily concerned with ensuring the health and safety of those involved in constructing, maintaining and cleaning a structure or building.

Duty holders need to be able to demonstrate competency in relation to health and safety standards and legal requirements.

What are the competency requirements for the Principal Contractor?

The Principal Contractor must plan, manage and monitor the building work and cooperate, coordinate and communicate to ensure the building work complies with the Building Regulations.

The Principal Contractor is expected to take responsibility for their site. Principal Contractors should manage the flow of information and make sure it gets to those who need it.

Principal Contractors are expected to work with designers and other contractors to make sure all building work complies with the relevant Building Regulations.

A Principal Contractor should have arrangements to monitor building work (and record this) to make sure it complies with the law and to liaise with the Principal Designer to agree any changes.

Summary of competence for all buildings

Legal and contractual requirements in relation to the discharge of their duties.

Ability to understand applicable legislation and recognise obligations regarding building safety.

Managing building work

Competency to manage building including knowledge of risks (fire and structural safety) and how to mitigate these.

Planning and organising production

The necessary competence to select competent contractors and suppliers. Competence to procure quality materials, products and building systems.

Managing construction processes/production

Ability to coordinate and monitor contractors, suppliers and service providers, specifically

- identification, assessment, inspection and testing of safety-critical materials, components and building systems
- ensuring any time or budgetary pressures do not impact building safety.

Leadership, decision making and change management

Competency to lead a building project using learning, experiences, and knowledge of standards and best practice. Principal Contractor will be able to take control of a build and manage change control but also delegate and empower others.

Liaising with the Client, other stakeholders and regulatory bodies

Develop a trusting, open and honest relationship with the key parties (Client, Principal Designer, Regulator).

Developing people and teams

Define the competencies required on a project and develop others to address any gaps.

Managing the quality of building work

Competencies to effectively manage build quality through communicating with and monitoring contractor's work. The Principal Contractor will also then need competencies to recognise and manage alterations or deviations.

Managing information

Competency to retain accurate and reliable documented information. In most cases, this will necessitate the competency to develop digital systems for the build information management.

Summary of competence for higher-risk buildings

The Principal Contractor shall meet the additional competences required to comply with the duties for higher-risk buildings. Specifically, the Principal Contractor shall have the skills, knowledge, experience and behaviour to

- contribute to the planning and development of the Construction Control Plan, understand the drivers of compliance with the applicable building regulations and record evidence of that compliance
- liaise with other duty holders, including the Client and Principal Designer, and understand the regulatory obligations (including building regulations) in respect of signing-off key information, including the compliance of the 'as-built' building
- coordinate matters and understand the drivers of compliance with regard to all those involved in the building work
- establish that all those working on the construction of higher-risk buildings, including contractors and building system providers, are aware of their duties and have the competence to carry out the construction activities for the safety of the building's occupants
- maintain the Construction Control Plan (which forms part of the golden thread of information) during the construction phase, detailing full and accurate records of all changes (including justifications) to the approved design
- establish an obligatory system for mandatory occurrence reporting on structural and fire safety, and other regulatory prescribed hazards, that enables the workforce to report potential occurrences which could cause a significant risk to life safety.

Persons appointing a Principal Contractor are recommended to consult document PAS 8672:2022: Framework for competence of individual Principal Contractors – Specification.

Are there any set procedures or templates that can be used to determine the competency of duty holders?

The Health and Safety Executive, via the Building Safety Regulator, recommends that clients, organisations and individuals follow the steps laid out in the PAS documents referred to above.

In addition, the Health and Safety Executive published a guidance document in 2022 – Building safety competence information for principal contractors and principal designers: Competency standard PAS 8671: 2022 and Competency standard PAS 8672: 2022.

While the core competencies of duty holders are likely to be the same for all projects, each project will have different needs and expectations which require a different set of competency criteria. So, one form or template will not necessarily suit all competency assessments.

A Client and any person appointing any person or organisation to work on a building project should establish their own criteria for competency, based on the PAS documents and any other recognised guidance documents, and should develop their own processes.

As an example, questionnaires can be used to obtain the basic information required, as has been in used for many years in relation to competency schemes required under CDM Regulations 2015 – for example, CHAS, SSIP, Construction Line – and this information can be supported by physical interviews of key personnel, interviews with previous clients, assessment of CPD materials, and so forth.

Whatever system is used to demonstrate competency, it is important that the process is recorded and written records kept.

Documented records can be used as evidence that everything reasonably expected to be done to check competency has been done and that the Client or person appointing the person to be involved in design or building works has not turned a blind eye to the requirements of the law.

Case study indicating lack of competency

Poorly applied external insulation on buildings (Report ID: 1203 Published: 21 August 2023; Region: CROSS-UK)

Overview

The report concerns external insulation on buildings being inadequately applied and fixed, with poor control of the workmanship standards, leading to potential safety concerns and damage to the property.

The reporter says workmanship issues observed include insulation applied in a haphazard method, multiple unnecessary holes drilled, insulation cut into shape on the installer's knee, and insulation panels having to be reattached several times because of defects in fixing. Generally, the reporter is concerned about poor standards of workmanship that can have safety implications and result in poor insulation for the buildings concerned.

The reporter considers there may be *poor training, a lack of appreciation of the risks of material failure, and a lack of checking of the standard of work.* The reporter wants to see higher standards of installation and inspection to avoid future failures and falls of material.

Key learning outcomes

For owners, clients and specifiers:

- Be aware that insulation and renders can fall if not properly designed and installed.
- Consider the advantages of work guarantee schemes that require providers to be competent.

For providers and installers of external insulation:

- External insulation fitted to buildings must be designed by a suitably qualified and experienced person (SQEP).
- Insulation installers should be appropriately trained and experienced.
- Installations should be monitored by competent supervisors.

The Construction Fixings Association provides guidance for those specifying, using and testing fixings.

PRINCIPAL DESIGNERS, DESIGNERS AND CONSULTANTS: COMPETENCY ASSESSMENTS (Example questionnaire)

Building Regulations etc. (Amendment) (England) Regulations 2023

Construction (Design & Management) Regulations 2015

Please add additional pages as and where necessary, stating the question number you are answering.

A. GENERAL INFORMATION

1. Name and Address of Organisation

   ....

   ....

   ....

   Post Code: ....

   Telephone Number: ....

2. Name of Head and Senior Person responsible for overseeing Building Regulations and CDM responsibilities

   ....

   ....

3. Professional Discipline(s)

   Please describe your principal activity, design discipline etc.

   ....

   ....

   ....

   ....

4. Membership of Professional Bodies

   a) Corporate Membership

   ....

   ....

   ....

   b) Individual(s) Membership

   ....

   ....

   ....

5. Please supply a copy of the organisation's Health and Safety Policy. Ensure specific areas as follows are addressed:

   - Health and safety arrangements
   - Health and safety responsibilities
   - Health and safety objectives
   - Health and safety risk assessment procedure
   - Health and safety training provisions
   - Appointment of competent person
   - Safety monitoring and auditing procedures
   - Health and safety procedures e.g. guidance on working practices etc.
   - Accident and incident reporting procedures

6. Please provide examples of accident and incident cases, accident statistics etc. and summaries of improvements you have made to internal procedures as a result of any investigations.

   ....

   ....

   ....

   ....

   ....

   ....

   ....

   ....

7. Please provide details of resources available for role as Principal Designer. Include estimated number of staff available for professional input, administrative input etc.

....

....

....

....

....

....

8. Please provide information on resources available in respect of legal and technical information in respect of understanding the requirements of the Building Regulations, fire safety and health and safety e.g. manuals, books, codes of practice, on-line databases, CD Rom, libraries, access to external consultants.

....

....

....

....

....

....

....

....

9. Please give any details of litigation, arbitration, criminal proceedings, issuing of stop notices, compliance notices in respect of building regulation compliance, health and safety, fire safety or environmental protection matters.

....

....

....

....

....

10. Please provide evidence of Professional Indemnity.

....

11. Do you have in-house or external safety expertise available to you?

    ....

    ....

    If external, please give name and address and supply details of any contractual arrangements.

    ....

    ....

    ....

    ....

    How often do you have access to this external advice?

    ....

12. Do you have in-house or external advice on construction methods, project management, construction project costings, resources and time management implications?

    ....

    ....

    ....

B. COMPETENCY, SKILLS AND TRAINING

13. Please provide specific qualifications, training, skills and experience of persons engaged as Principal Designers on this project.

    ....

    ....

    ....

    ....

    ....

14. Do any of your employees hold any independently awarded safety accreditation cards e.g. CSCS Professional Card or other European or international cards etc.?

    YES/NO

    If yes, please describe

    ....

    ....

15. What procedures do you have in place to ensure that all your designers and other employees are kept up to date with current trends in health and safety, design solutions, construction methods etc.?

....

....

....

....

16. Who, within your practice, manages your CPD procedures and records?

....

....

17. How do you formally assess each individual's attainment of their various CPD requirements?

....

....

....

....

18a. What formal training courses do you provide for employees, especially in relation to health and safety and new developments in building regulations and other regulatory matters?

....

....

....

....

18b. Please describe the training your employees have for asbestos awareness as required by the Control of Asbestos Regulations 2012.

....

....

....

....

C. CDM MANAGEMENT PROCEDURES

19. Do you have a CDM in-house manual? If yes, please attach contents page.

    ....

20. Has your organisation ever been appointed as Principal Designer (CDM) to any project? If yes, please give examples.

    ....

    ....

    ....

    ....

    ....

    ....

21. Has your organisation developed pro-formas for design hazard and risk assessments? If yes, please attach examples. If no, please describe how you ensure relevant consideration is given to health and safety matters on projects and how information is recorded and shared amongst those who need to know it.

    ....

    ....

    ....

    ....

22. What record format is in use to detail considerations given to health and safety issues on the project e.g. written records, CAD information, risk assessment forms, computer records etc.?

    ....

    ....

    ....

    ....

    ....

23. How have your design techniques and procedures improved so as to accommodate the specific health and safety responsibilities given to Designers under the CDM Regulations 2015? In particular, how will you be ensuring compliance with the Workplace (Health, Safety and Welfare) Regulations 1992 and ensuring that consideration is given to the principles of prevention as referenced in CDM 2015?

....

....

....

....

24. What in-house quality assurance, total quality management or other auditing procedures do you use to demonstrate that formal consideration is given to differing project stages and that your duties as designers under the CDM Regulations 2015 are being met?

....

....

....

....

....

....

D. DESIGN RISK ASSESSMENTS

25. Please indicate the main design issues you have to deal with in relation to managing health and safety on projects e.g. specification of materials, sequencing construction activities, structural engineering issues e.g. steelworks, excavations etc.

....

....

....

....

....

26. Please give examples of how you have, as a design practice, designed out or reduced the residual risks from hazards associated with your design.

....

....

....

....

27. Do you prepare Design Risk Registers or subscribe to the management control process of Red, Amber, Green risks as referenced on the HSE website? YES/NO

If yes, please give examples:

....

....

....

....

28. If your design solutions create residual hazards, how do you ensure that the Client or end user of the building is fully briefed on those hazards and what type of information do you provide as an indication of suitable control measures to reduce the risks from the residual hazards?

....

....

....

....

29. How do you provide information for inclusion in the Health and Safety File and for inclusion in the Safety Case information for higher-risk buildings?

....

....

....

....

E. REFERENCES

30. Please supply the names and contact details of two references – preferably current and former Clients.

    i)

    ....

    ....

Relationship: ....

    ii)

    ....

    ....

Relationship: ....

Form completed by (please print name): ....

Position in company: [Database] ....

Date: ....

Signed: ....

The Building Safety Act 2022: Questions and Answers: A practical approach to design and management of higher-risk buildings

Pat Perry
ISBN 978-1-83662-467-7
https://doi.org/10.1108/978-1-83662-464-620251006

Chapter 6
The Building Safety Regulator

What is the Building Safety Regulator?

A key recommendation of Dame Judith Hackitt's 2018 review of the building industry was the creation of a new national regulator, tasked with promoting and enforcing safety in the built environment. That recommendation has become a cornerstone of the Building Safety Act 2022.

The Building Safety Act names the Health and Safety Executive as the new Regulator and a new division has been created within the Executive to manage and enforce the legal requirements.

Under the Building Safety Act, the Regulator has three main functions

- overseeing the safety and standards of all buildings
- helping and encouraging the built environment industry and building control professionals to improve their competence
- leading the implementation of the new regulatory framework for higher-risk buildings.

What is the purpose of the Building Safety Regulator?

In the Building Safety Regulator's strategic plan for 2023–2026, it states that its mission is 'protecting people and places'.

Its vision is to create a built environment where everyone is competent and takes responsibility to ensure buildings are of high quality and safe, so residents and other building users are confident that the tragedies of the past will never be repeated.

It states

> We want to:
>
> - improve the safety and standards of all buildings;
> - make sure residents of higher-risk buildings are safe and feel safe where they live; and
> - help restore trust in the built environment sector.
>
> We will do this by:
>
> - delivering consistent standards within the building control profession;
> - overseeing and driving improvements across the built environment;
> - regulating the planning, design and construction of new higher-risk buildings;

- ensuring those who are responsible for higher-risk buildings manage risk so that residents are safe; and
- working in partnership with co-regulators.

What is the most important function of the Building Safety Regulator?

The Building Safety Act 2022 introduced a new stricter regime for the regulation of higher-risk buildings in order to ensure that proper scrutiny of plans and building proposals were undertaken by a specialised and competent body so as to ensure that due regard is paid to fire safety and structural safety matters alongside adequate operational and management procedures for the subsequent occupied building.

Part 3 of the Building Safety Act defines higher-risk buildings as those that are at least 18 m in height, or with at least seven storeys, and that contain two or more dwellings.

The Higher-Risk Buildings (Descriptions and Supplementary Provisions) Regulations 2023 extend that definition by providing that care homes and hospitals also fall within the definition of higher-risk buildings (if they meet the height/number of storeys requirements in the Building Safety Act).

For higher-risk buildings, the new Regulator alone will enforce the provisions of the Building Safety Act and any regulations thereunder.

Is the Building Safety Regulator the Building Control Authority for higher-risk buildings?

Yes. The Building Safety Regulator became the Building Control Authority for all higher-risk buildings in England in October 2023.

Developers need to apply to the Building Safety Regulator for building control approval for all new building work in-scope of the higher-risk regime. This work can no longer be overseen by local authority or private sector building control.

Anyone applying to the Building Safety Regulator for building control approval needs to ensure that their applications clearly and comprehensively demonstrate compliance with the Building Regulations.

The Grenfell Tower Public Inquiry, chaired by Sir Martin Moore-Bick, concluded in its Final Report about the insulation material used on the Tower

> The LABC [Local Authority Building Control] was vulnerable to manipulation because its processes were not implemented rigorously enough. The task of producing an initial assessment should not have been given to building control officers, who did not have the degree of knowledge and experience necessary to make an informed assessment of the product in question, and those who carried out the second stage review were not always competent to do so and in some cases did not take the necessary degree of care.
>
> Over a period of some years the LABC's certificates relating to Kingspan K15 and Celotex RS5000 contained misleading statements about their fire performance and about the suitability

of both products for use in the external walls of buildings over 18 metres in height. Despite warnings from various quarters, the LABC failed to scrutinise properly the claims made for the products by the manufacturers and instead adopted uncritically the language they suggested. In short, it was willing to accommodate the customer at the expense of those who relied on the certificates. As a result, the LABC was also the victim of dishonest behaviour on the part of unscrupulous manufacturers.

The Building Safety Regulator's role is to act as a regulatory authority, not a service provider, meaning it verifies compliance without offering consultation services.

Does the Building Safety Regulator oversee the building control requirements for refurbishment works to existing higher-risk buildings?

Yes.

Alterations made to existing higher-risk buildings such as refurbishment works to an existing flat in a higher-risk building require the Building Safety Regulator's approval.

Applicants must demonstrate that their proposed alterations will comply with all relevant Building Regulations and that they will be adequately managed during construction.

Works to a building that are intended to convert an existing non higher-risk building into a higher-risk building fall within scope of the new building safety regime and a building control application must be made to the Building Safety Regulator – for example, the addition of a storey to an existing six storey building thus making it seven storeys and higher than 18 m or the conversion of an office block to residential accommodation.

The control of refurbishment works to higher-risk buildings is seen as critical to the overall safety of buildings, as it was the refurbishment works to Grenfell Tower that were not adequately controlled nor properly scrutinised by any of the regulatory bodies nor by the Client commissioning the works.

For all works to existing higher-risk buildings that are not exempt, landlords must follow the gateway regime under the Procedures. Failure to meet the requirements will be an offence under the Building Act 1984.

All higher-risk buildings will be subject to Part 3 of the Building Safety Act, which imposes new obligations in respect of the design, construction and refurbishment of those buildings.

Are there any specific Regulations setting out the procedures that need to be followed when applying for building regulation approval for either new building works or works to existing buildings?

In August 2023, the government published the Building (Higher-Risk Buildings Procedures) (England) Regulations 2023 (the 'HRB Regulatory Regulations'), which implement Part 3 of the Building Safety Act.

The HRB Regulatory Regulations set out the details of the building control regime for higher-risk buildings, specifying the procedural building regulation requirements when a new higher-risk building is being designed and constructed or when building work is being done to an existing higher-risk building. The new regime came fully into force in April 2024.

What is the Building Safety Regulator's approval role?

The Regulator will be involved throughout the design and construction of a new higher-risk building, or when work is being done to an existing higher-risk building or, if work is being done to an existing building, following the completion of which that building will fall within the definition of a higher-risk building.

The Regulator is required to approve applications before work starts (known as gateway two) within 12 weeks of receiving a valid application (see Chapter 3 on gateways).

The Regulator can require a Client to notify it when a specified point of the work has been reached and not to cover up specified work for a specified period.

Certain changes must be notified to the Regulator before works start ('notifiable changes') while the most significant changes ('major changes') must be approved by the Regulator before works can start.

The Regulator has 6 weeks from receiving a valid application for building control approval to approve a major change. Where the Regulator does not respond within the requisite time (as may be extended by written agreement), the Regulations do not provide for approval by default, meaning that a contractor will have to wait until it obtains the Regulator's approval before it is able to start works in respect of the major change.

Where, during the construction phase, (*i*) an aspect of the design relating to the structural integrity or fire safety of a higher-risk building would give rise to a risk of a significant number of deaths, or serious injury to a significant number of people, or (*ii*) an incident or situation relating to the structural integrity or fire safety of a higher-risk building occurs that would be likely to present a risk of a significant number of deaths, or serious injury to a significant number of people, then a principal duty holder must, on becoming aware of the occurrence, notify the Regulator of the safety occurrence by the quickest practicable means without undue delay and then provide a written report to the Regulator within 10 days. This is known as the Mandatory Occurrence Notice and Report.

The Regulator will issue a Completion Certificate in respect of work to a higher-risk building (known as gateway three) within 8 weeks of receiving a valid Completion Certificate application.

The Regulator also has a role in approving applications for work at the various stages for design and construction.

What is the Regulator's enforcement role?

In addition to its role in approving works, in relation to any building or proposed building for which the Regulator is the Building Control Authority, the Regulator may take such steps as it considers appropriate to check compliance with all applicable requirements of the Building Regulations, including

- requiring information
- requiring the laying open of building work for inspection by the Regulator
- requiring an inspection to be undertaken and the record of the inspection provided.

There are also two forms of notice available to the Regulator

- compliance notices: if the Regulator considers that a person appears to have contravened, be contravening or be likely to contravene the Building Regulations, they can issue a compliance notice, which effectively requires the recipient to take specified steps within a specified period (to ensure compliance)
- stop notices: these prohibit, either immediately or from a specified time, the carrying out of specified work. They can be issued to a person who appears to be in control of any work if it appears that it
    - contravenes specific Building Regulations
    - contravenes a Compliance Notice; or
    - contravenes a building regulation which can cause serious harm – meaning that there is a risk of serious harm to people in or around the building if the contravention is not corrected.

A person who, without reasonable excuse, contravenes these notices commits a criminal offence, punishable by a fine and/or up to 2 years in prison.

The Building Safety Regulator has published an Enforcement Policy Statement that sets out the key enforcement principles in respect of the Regulator's enforcement powers, regulatory tools and sanctions, and how they will be applied. The principles are

- proportionality in how the Regulator applies the law and secures compliance
- targeting of enforcement action
- consistency of the Regulator's enforcement approach
- transparency about how the Regulator operates and what stakeholders can expect
- accountability for its actions.

What other duties does the Building Safety Regulator have?

The Building Safety Regulator regulates registered building inspectors and building control bodies, including

- local authorities
- registered building control approvers.

Businesses that carry out building control work must be registered by 6 April 2024 to operate as a registered building control approver (RBCA).

Building inspectors/building control officers must be registered with the Building Safety Regulator to carry out Building Safety Regulator-regulated activities such as assessing plans and inspecting. This includes all buildings from domestic premises to complex, higher-risk, multi-use structures.

Why do individuals and organisations carrying out building control works have to be registered?

The Public Inquiry reported as follows

> The evidence shows that in the period leading up to the Grenfell Tower fire many of those involved in major construction projects, including clients, contractors and even architects, regarded building control primarily as a source of advice and assistance. It was even described as an extension of the design team. In many cases that was how building control itself saw its role. That was a serious misunderstanding, but it was fostered by building control bodies themselves, who preferred to co-operate with applicants to enable proposals to be approved rather than enforce the Building Regulations rigorously. In our view, that has to change.
>
> The government has taken steps to improve the regulation of building control and the competence of those who consider applications for approval. We expect the construction regulator to continue these new arrangements, which are intended to introduce a wholly new climate in which both applicants for approval and building control officers understand that the function of building control is regulatory in nature.

Competency standards of several professional roles were reviewed by the Inquiry and generally, it was found that there were inconsistencies in the approach to regulation and that professional standards and knowledge were lacking.

The new registration scheme will ensure that only those with the appropriate level of knowledge, experience and skill can practise as a building inspector as they will need to re-register every 4 years, thus demonstrating that they have kept their knowledge up to date.

After April 2024, anyone who intends to carry out

- assessing plans
- inspections
- giving advice to building control bodies that carry out regulated functions

in England or Wales, must be registered as a building inspector. After this date, anyone not registered is committing a criminal offence to work as a building inspector.

The Building Inspector Competence Framework (BICoF) has been introduced and provides a clear, structured framework to demonstrate competence for the whole building control profession.

Competence reflects the necessary skills, knowledge, experience and behaviours of individuals performing their role as a registered building inspector.

What is the role of multi-disciplinary teams within the Building Safety Regulator's department?

The Building Safety Regulator will bring together experts such as building inspectors, structural engineers and fire safety professionals when necessary to effectively manage the safety of higher-risk buildings. They provide regulatory advice, conduct inspections, and ensure effective decision-making and compliance with the Building Regulations for higher-risk buildings. Multi-disciplinary teams (MDTs) guide the structure of a building inspection schedule and input to any requirements or conditions that may be applied to a building control approval.

Each higher-risk building project is assigned a dedicated MDT, led by the Building Safety Regulator.

They also provide a main point of contact for applicants, and they review Completion Certificate applications with the 'as-built' information to ensure compliance with the Building Regulations before issuing the Completion Certificate. This is not new in terms of assessments and site inspections – but MDTs are a new method of delivery.

MDTs work with a structured inspection schedule, ensuring that safety and compliance are continually checked throughout the project life cycle.

What are the responsibilities of multi-disciplinary teams?

Once a building control project begins, the MDT assigned to the project is responsible for the following.

- Assessing applications: the team reviews the initial building control application and works together to assess all submitted plans.
- Setting an inspection schedule: MDTs agree an inspection schedule focused on compliance with the functional requirements of the Building Regulations throughout the building's construction.
- Inspections and interventions: the team conducts inspections to ensure that the construction meets the required safety and standards and takes immediate action if any issues arise.
- Ongoing communication: the MDT provides a lead contact for the applicant, enabling communication throughout the process.
- Completion Certificate review: at the end of the project, the MDT reviews the 'as-built' documentation to confirm that the finished building meets all requirements before issuing a Completion Certificate.

The Public Inquiry raised concerns about the competency, cooperation and communication processes adopted within higher-risk building projects and concluded that many organisations were incompetent.

The idea of the MDT within the Building Safety Regulator is designed to reduce failures and oversights of critical design and build stages involving key matters of safety.

The use of MDTs marks a significant shift in how building safety and standards are regulated. By ensuring that all necessary expertise is involved from the outset, this method seeks to prevent dangerous oversights and improve overall safety and standards for higher-risk buildings in England. BSR is committed to leading this change and ensuring that the safety of residents remains a top priority. (The critical role of Multi-Disciplinary Teams in building safety and standards – Gov.uk)

Does the Building Safety Regulator charge for its services?

Yes, it can do.

The Building Safety (Regulator's Charges) Regulations 2023 enables the Building Safety Regulator to charge for certain functions under the Building Act 1984 (as amended) and the Building Safety Act 2022.

In general terms, applications for building control approval are chargeable, with both an application fee and hourly charges of approximately £151/h (as of April 2025) for Regulator staff.

In respect of higher-risk buildings, the Regulator can charge for

- applications and registrations of higher-risk buildings
- inspections
- reviews and appeals.

Again, costs will be fixed fees for applications – for example, Completion Certificates, higher-risk building registrations – and hourly fees for any further work needed by the Regulator.

If a person/company is thought to be contravening the legislation and the Regulator decides to investigate and take formal action to ensure compliance, there will be a charge for such intervention. This is similar to the existing 'fee for fault' charged by the Health and Safety Executive for health and safety investigations.

REFERENCE

Building Safety Act 2022.
Building Safety Regulator Strategic Plan 2023–2026
Building Safety (Regulator's Charges) Regulations 2023

The Building Safety Act 2022: Questions and Answers: A practical approach to design and management of higher-risk buildings

emerald PUBLISHING ice Publishing

Pat Perry

ISBN 978-1-83662-467-7
https://doi.org/10.1108/978-1-83662-464-620251007

Chapter 7
The Client

What is meant by the term 'Client' under the Building Safety Act 2022?

The Client is the organisation or individual for whom a construction project is carried out. They are ultimately responsible for ensuring that the requirements of the Building Safety Act 2022 and the Building Regulations are satisfied.

The Client's duties cannot be subcontracted to a third party.

What is the essential duty of the Client?

As the Client, you must make suitable arrangements for planning, managing and monitoring a project so it complies with all relevant requirements on completion.

In particular, a Client must

- allocate enough time and resource for the building work to comply with the Building Regulations
- establish, review and maintain systems and arrangements to meet the Building Regulations
- cooperate with others working on the project so they can comply with their duties
- enable cooperation between designers and contractors.

As the Client, you must provide building information to every designer and contractor working on the project. Building information relates to

- the building work or design work
- the planning and management of the project
- any issues involving compliance with relevant requirements and how they have been addressed.

I understand that I must appoint duty holders if there is or is likely to be more than one contractor on site. Surely these duty holders will be responsible for making suitable arrangements to manage and supervise the project?

Clients commonly believe that appointing a Building Regulations Principal Designer and a Building Regulations Principal Contractor is enough to discharge their responsibilities under the Building Regulations, but this is not the case.

Clients must work with their duty holders and collaborate regarding sharing information and taking responsibility for the successful and legal completion of the building project.

In summary, what are the Client's responsibilities under the Building Safety Act 2022?

It is the Client's responsibility to

- ensure design work is carried out so that, if built, the building work would be compliant with the Building Regulations
- ensure the building work is carried out in accordance with the Building Regulations
- enable designers and contractors to cooperate to ensure compliance with the Building Regulations
- undertake periodic reviews to identify any higher-risk buildings
- provide building information to every designer and contractor on the project
- ensure arrangements are maintained and reviewed throughout the project
- assess the competency of the Principal Designer and Principal Contractor
- make Client appointments in writing as follows if there is more than one contractor
    - a Building Regulations Principal Designer
    - a Building Regulations Principal Contractor.

When do appointments need to be made?

Appointments must be made before the construction phase begins or, if the project involves a higher-risk building, before a building control application is submitted to the Building Safety Regulator.

What happens if the Client fails to make these appointments?

If the Client does not appoint a Building Regulations Principal Designer or Building Regulations Principal Contractor, they must fulfil the duties of those roles.

I have already appointed a Principal Designer and a Principal Contractor under the Construction (Design and Management) (CDM) Regulations 2015. Do I need to make additional appointments?

Where a Client has already appointed a Principal Designer and Principal Contractor under the CDM Regulations 2015 for the project, the Client may certify in writing that those parties will also be appointed as the Building Regulations Principal Designer and Principal Contractor under the Building Safety Act.

The Building Safety Act 2022 contains many references to duty holders and their competence. What does this mean for the Client?

You must take all reasonable steps to appoint designers and contractors with the necessary competence or organisational capability to carry out their roles.

Clients themselves do not need to demonstrate any levels of competency unless they are taking on the roles of Principal Designer and Principal Contractor.

Clients will need to be able to demonstrate that they have procedures in place to assess the competency of the duty holders they appoint and will need to understand the level of competency their project will demand. It will not be appropriate for an organisation or individual to be appointed as a duty holder to a project involving a higher-risk building if they have no experience in working on similar projects.

While there is no express statutory requirement that clients keep written records of these appointments, it will be prudent for a Client to keep such records in order to be able to prove that they have properly investigated the competency requirements before an appointment in writing is made. It will also be sensible that appointments include a representation from the relevant appointees that they meet the competency requirements.

What additional requirements relating to competency are placed on clients in respect of higher-risk buildings?

In respect of higher-risk building works (as defined in s.120D of the Building Act 1984 (as amended by the Building Safety Act) and the Higher-Risk Buildings (Description and Supplementary Provisions) Regulations 2023)) on each appointment of a Principal Designer or a Principal Contractor, the Client must keep a record, in writing, of the reasonable steps they took to satisfy themselves that the Principal Designer and Principal Contractor meet the competence requirements.

Evidence helps demonstrate that, as the Client, you did everything reasonably practicable to comply with your duties. Any investigation of an on-site accident, for example, would more than likely involve the investigating officer to review and assess whether the designer or contractor knew what they were doing and whether they were aware of any hazards and risks associated with the task – that is, did they have the experience, knowledge and skill, and where appropriate, the behaviours, necessary to carry out the task safely?

The term 'serious sanctions' is used in the Building Safety Act 2022 and its secondary legislation relating to higher-risk buildings. What does the term mean?

Where the work relates to a higher-risk building, the person appointing a duty holder must ask the appointee whether a serious sanction has occurred, in relation to them, within the 5 years ending on the date of the appointment and consider any information available relating to any misconduct (including any serious sanction).

A serious sanction can be

- a compliance notice that refers to a contravention or likely contravention of a requirement of Part A (structure) or Part B (fire safety) of Schedule 1 of the Building Act 1984
- a stop notice under the Building Act 1984
- convictions under the Building Act 1984, the Health and Safety at Work etc. Act 1974, the Building Safety Act 2022, or the Regulatory Reform (Fire Safety) Order 2005.

A serious sanction can also be a report published under the Inquiries Act 2005 that finds an appointee's actions, or inaction, resulted in one of the following

- one or more deaths
- a likely contravention of any requirement under the Building Act 1984, the Health and Safety at Work etc. Act 1974, the Building Safety Act 2022, or the Regulatory Reform (Fire Safety) Order 2005.

You must consider how any serious sanction and any resulting action by the individual or organisation affects their judgement or their capability to do the work.

Where an individual or organisation has declared that they have had a serious sanction against them, it may not always be appropriate to decline their appointment. The information you want to obtain is: what did they learn from the experience and what new controls and changes to procedures, training, competency skills, and so forth, have they introduced?

Accidents do happen and when they do, it is essential to look for positive outcomes and to investigate how they have improved the health and safety management process or fire safety procedures.

Should you decide to appoint a duty holder with a serious sanction history, it will be essential to record the reasons why you appointed them, the evidence you reviewed to give confidence that lessons have been learned and what, if any, additional monitoring you will put in place while they undertake their duties.

What steps need to be taken if there is more than one Client on a building project – for example, a mixed-use development with residential, commercial and retail development schemes?

When there is more than one Client for a project, the clients can nominate a lead Client. All clients must confirm this in writing.

When there is a lead Client, they must carry out all the Client's duties for the project.

The other clients, who are not the lead Client, must

- provide building information to every designer and contractor working on the project
- cooperate with others working on the project to enable them to comply with their duties
- allow for a periodic review of the building and design work to identify whether it is higher-risk building work
- provide information to designers and contractors so they are aware the project involves higher-risk building work.

Are there any duties placed on a person who commissions a domestic project for their own use – that is, someone who is a domestic Client under the legislation?

Domestic clients have been defined as 'for whom a project is being carried out which is not in the course or furtherance of a business of that client'. For example, a homeowner wishing to have an extension, conversion or alteration to their property would fall into this category.

Their key responsibilities are to

- provide building information as soon as is practicable to every designer and contractor on the project. This information could include existing services or infrastructure (electrical, gas or water), potential contamination or historic usage that would likely have an impact on the project
- cooperate with any other person working on the project to enable any person with a duty or function under the Regulations to fulfil that duty or function
- in circumstances where there is more than one Client involved in a project, obtain an agreement in writing to identify who is responsible for the duties owed by a Client under the Regulations
- provide the necessary information to all designers and contractors working on projects considered higher-risk building work and carry out a periodic review to establish if/or when the work becomes higher-risk work
- appoint a contractor, who on their behalf must carry out the duties to ensure that
    - suitable arrangements are made for planning, managing and monitoring a project, including the allocation of sufficient time and other resources
    - design work, if built, would comply with all relevant requirements
    - building work is carried out in accordance with all relevant requirements
    - cooperation between the designers and contractors is enabled to ensure compliance is achieved.

Does a Client have any additional duties in respect of higher-risk buildings?

Yes, the Client does have some extra duties in relation to higher-risk building, especially around the requirement to provide information.

Regulation 11(B) of the Building Regulations etc. (Amendment) (England) Regulations 2023 states

> A client must make suitable arrangements to ensure information is provided to the designers and contractors working on a project which includes any higher-risk building work to make them aware that the project includes higher-risk building work and the nature of the higher-risk building work.
>
> Paragraph (1) includes a duty to periodically review the building work (and the design work) included or to be included in the project so as to identify whether it is higher-risk building work and to ensure information is provided under paragraph (1) where the work becomes higher-risk building work.

Sharing information is a critical stage in ensuring that everyone working on a project knows whether they are dealing with a higher-risk building and the Client is expected to take the lead in initiating an information sharing culture.

Can clients delegate their legal responsibilities to other duty holders?

No. While a Client can delegate tasks to other people or organisations, they cannot delegate their legal responsibilities and will always be held accountable for the duties allocated to clients.

Clients may have a defence against any criminal charges if they can prove that they did everything reasonably practicable to ensure that anyone they delegated tasks to was competent to carry out those tasks and that they had systems in place to check progress and compliance during the project.

Health and safety law together with general business law allows for senior managers within a body corporate to be prosecuted for offences if it can be shown that the offence occurred due to the 'consent or connivance' of the individual concerned.

In Huckerby v Elliot [1970], the Divisional Court stated that a person is said to have connived at an offence when

> … he is equally well aware of what is going on but his agreement is tacit, not actively encouraging what happens but letting it continue and saying nothing about it.

Connivance therefore encompasses wilful blindness – that is, 'turning a blind eye'. Clients cannot abdicate their legal responsibilities by assuming that they do not have to take any interest in the project and that others will 'carry the can'.

The Building Safety Act 2022: Questions and Answers: A practical approach to design and management of higher-risk buildings

Pat Perry
ISBN 978-1-83662-467-7
https://doi.org/10.1108/978-1-83662-464-620251009

Chapter 8
Accountable Persons

New roles of 'Accountable Person' and 'Principal Accountable Person' have been created under the Building Safety Act 2022. Who are they and what are their duties?

The Accountable Person is the entity responsible for meeting the statutory obligations for occupied higher-risk buildings. The Accountable Person will usually be the person who either has the legal estate in possession of, or is under a relevant repairing obligation for, any part of the common parts of the building.

An Accountable Person is an organisation or individual who owns or has a legal obligation to repair any common parts of the building.

Common parts are used by residents, such as

- the structure and exterior of the building
- corridors
- lobbies
- staircases.

An Accountable Person can be a

- freeholder or estate owner
- landlord
- management company
- resident management company
- right to manage company
- commonhold association.

Where there are two or more persons with legal interest in a building in scope, the Accountable Person will be the one responsible for the repair and maintenance of any common parts of the building.

Where there are two or more Accountable Persons for a building, the person/organisation responsible for the exterior and structure of the building will be designated 'Principal Accountable Person'.

This is the legal person who under the new regime will be responsible for the fire and structural safety of a 'higher-risk' building.

The Accountable Person will be responsible for assessing building safety risks on an ongoing basis, and for taking all reasonable steps to manage those risks by preparing a Safety Case Report that must be submitted to the Building Safety Regulator.

New criminal sanctions (including imprisonment up to 2 years and unlimited fines) will apply to the Accountable Person. Criminal liability could also attach to individual directors, managers and members of the Accountable Person, if it is a body corporate, where there is consent, connivance or neglect by the individual.

The Accountable Person could be the building owner, the managing agents or any individual or corporate entity that has repairing obligations for the building.

Accountable Persons and the Principal Accountable Person cannot delegate their legal obligations to others. Accountable Persons and the Principal Accountable Person can employ an organisation or individual, like a managing agent, to carry out duties on their behalf. However, the accountability for making sure those duties are carried out and the liability for a building's safety remains with the Accountable Persons and Principal Accountable Person.

What are the duties of the Principal Accountable Person and the Accountable Person?

All Accountable Persons must ensure that they have effective, proportionate measures in place to manage building safety risks in the buildings or parts of the building that they are responsible for. In buildings with multiple Accountable Persons, they must – so far as possible – cooperate and coordinate with all other Accountable Persons to deliver building safety to the residents.

The core duties of Accountable Persons revolve around the assessment and management of building safety risks in their higher-risk building.

All Accountable Persons have a duty to take all reasonable steps to

- put measures in place to prevent building safety risks happening and reduce the severity of any incident that does happen
- report certain fire and structural safety issues or incidents
- engage with residents about the building's safety and perform duties relating to the residents' engagement strategy
- keep, update and provide information about the building for the building's Safety Case
- transfer building safety information to any incoming Accountable Person
- notify the Building Safety Regulator if there is a change to an Accountable Person.

In addition to assessing and managing building safety risks in their own area of the building, the Principle Accountable Person has additional responsibilities for meeting specific statutory obligations for the building as a whole. They must

- apply for a Building Assessment Certificate when directed by the regulator and display it in the building
- display a notice in the building information about the Accountable Persons for the building, including their full contact details and a summary of their roles and duties
- prepare a Safety Case Report containing an assessment of building safety risks carried out by the Accountable Persons, and steps they have taken to address those risks or reduce the severity of incidents arising from them
- prepare a residents' engagement strategy, setting out how residents will be involved in building safety decisions
- establish and operate a mandatory occurrence reporting system to enable information on building safety risks to be captured and reported to the Building Safety Regulator
- set up and operate a system to investigate relevant complaints relating to building safety risks or the performance of an Accountable Person's duties
- keep, update and provide information about the building for the building's safety case.

Accountable Persons must follow the 'prescribed principles' when managing risks in higher-risk buildings. What are these?

The prescribed principles follow broadly the hierarchy of risk control, familiar to many under the Management of Health and Safety at Work Regulations 1999.

The Higher-Risk Buildings (Management of Safety Risks etc.) (England) Regulations 2023 set out the prescribed principles, in Regulation 4, required by section 84 Building Safety Act 2022, as follows

(a) avoid building safety risks (as defined in section 62 Building Safety Act 2022)
(b) evaluate building safety risks that cannot be avoided, including identifying the proportionate measures required to address, reduce, mitigate and control the risks
(c) combat building safety risks at source by introducing proportionate measures to address, reduce, mitigate and control the risk at the earliest opportunity
(d) ensure suitable and proportionate systems are in place for the effective inspection, testing and maintenance of the efficacy of measures taken
(e) give collective protective measures priority over individual protective measures
(f) adapt to technical progress
(g) where reasonable to do so, replace the dangerous with the non-dangerous or less dangerous
(h) consider the impacts on residents and carry out engagement with residents
(i) give appropriate instructions and information to employees and persons working on or in the building
(j) ensure that any person responsible for or assisting with compliance with a duty in or under Part 4 of the 2022 Act has the relevant competence.

What are building safety risks as defined in the Building Safety Act?

A 'building safety risk' is defined as any risk to the safety of persons in or about a building arising from fire, structural failure or any other matter prescribed in regulations.

The clause is drafted to capture risks inside and outside the building, such as fire spreading on the external wall of a building, but is limited to risks that are associated with the building and/or arising from the building itself. Risks are also limited to those affecting people in and about the building.

The explanatory notes to the Act state that the scope is only intended to encompass major and rapid onset events and that the government 'has no current plans for the higher-risk building regime to extend beyond the risks of fire and structural integrity'.

However, the Secretary of State can request advice from the Building Safety Regulator about new and emerging risks and may introduce additional definitions for building risk.

Statutory guidance may also be issued to help determine building risk and may include electrical and gas failures as a building safety risk, integrity of windows and glazing due to high winds, and so forth.

What is a 'higher-risk building' under the Building Safety Act?

Exact wording in the Act states that

> "higher-risk building" means a building in England that—
>
> (a) is at least 18 metres in height or has at least 7 storeys, and
> (b) contains at least 2 residential units.

The definition in the Act of residential unit is

> "residential unit" means—
>
> (a) a dwelling, or
> (b) any other unit of living accommodation;

Therefore, in addition to residential buildings, care homes and hospitals can also be considered higher-risk buildings if they are higher than 18 m or have at least seven storeys and so would fall under the new regulatory regime.

Regulation 2 of the Higher-Risk Buildings (Descriptions and Supplementary Provisions) Regulations 2023 sets out the buildings which are included in the design and construction part of the higher-risk regime as being at least 18 m in height or having at least seven storeys and contains at least two residential units or is a care home or is a hospital.

Section 65 of the Building Safety Act 2022 sets out the buildings which are included in the occupation part of the higher-risk regime. These are buildings of at least 18 m in height or having at least seven storeys and contain at least two residential units.

Regulation 8 of the Higher-Risk Buildings (Descriptions and Supplementary Provisions) Regulations 2023 sets out the buildings that are excluded from the occupation part of the higher risk regime

- hospital
- care home
- secure residential institution
- hotel
- military barracks and living accommodation for military personnel
- prison.

All higher-risk buildings must be registered with the Building Safety Regulator either before occupation of a new build or by 30 September 2023 if an existing building in occupation.

Who or what is 'a single point of contact'?

If the Principal Accountable Person is an organisation, then someone from the organisation can be the single point of contact for the Building Safety Regulator.

The single point of contact can also be a third party separate to the organisation, such as a management company. This individual should have authority or duties relating to the safety of the building, but this does not make them or the third party the Principal Accountable Person. It is the organisation that is the Principal Accountable Person.

The Principal Accountable Person must give written consent to the third party. If this consent stops, the Principal Accountable Person must update the single point of contact with the Building Safety Regulator.

What are the penalties for not complying with the duties imposed by the Act?

The Building Safety Regulator will have powers to prosecute all offences in Parts 2 and 4 of the Act (including Schedule 2), and the Building Act 1984. In addition, for all offences in the Building Act 1984 and Parts 2 and 4 of the Building Safety Act 2022, where an offence is committed by a corporate body with the consent or connivance of a director, manager, and so forth, of that corporate body, or is attributable to their neglect, that person will be liable to be prosecuted as well as the corporate body.

The Building Safety Regulator will be able to issue compliance notices (requiring issues of non-compliance to be rectified by a set date) and in design and construction, stop notices (requiring work to be halted until serious non-compliance is addressed).

Failure to comply with compliance and stop notices will be a criminal offence, with a maximum penalty of up to 2 years in prison and an unlimited fine.

The Act includes powers of entry to gather evidence for compliance action. A warrant from a magistrates' court will be required for premises used wholly or mainly as a private dwelling, or where force needs to be used to enter any premises.

The Building Safety Regulator will hold to account local authorities, registered building control approvers and registered building inspectors – for example, where they have not registered or are performing below the set standard – and will be able to suspend or remove inspectors from the register and to prosecute where necessary.

The Act also includes powers for the Building Safety Regulator to apply to the First-tier Tribunal for an order appointing a Special Measures Manager to take over the functions of the Accountable Person where there is a significant failure or repeated failures to comply with their statutory duties.

Accountable Persons are responsible for registering a building with the Building Safety Regulator. What does this mean?

High-rise residential buildings must be registered with the Building Safety Regulator.

It is an offence to allow residents to occupy an unregistered high-rise residential building.

A high-rise residential structure has

- at least seven floors or is at least 18 m in height
- at least two residential units.

A high-rise residential building can contain one or more high-rise residential structures.

Apply to register two or more structures as a single building when they are connected by either

- a walkway, lobby or basement that contains a residential unit
- an internal wall containing normal-use doors.

You must give the Building Safety Regulator the following information about the building

- the number of floors at or above ground level
- its height in metres
- the number of residential units
- its completion date
- its address or addresses
- the names and contact details of the Principal and Accountable Person
- the name of the single point of contact
- structure and fire safety information – that is, key building information such as: fire and smoke controls, energy supplies, storage and generation, type of structure, roof, staircases and external walls, the building's use, building work since original build, connections between structures or to other buildings.

If your building is made up of more than one high-rise residential structure, you must give the Building Safety Regulator this information for each structure.

Registration can take place online. There is a fee of £251 per building (April 2025).

Required building information when registering a high-rise residential building

Fire and smoke controls

What strategy do you use to evacuate the residential part of this building or structure?

Evacuation strategies may be

- phased
- progressive horizontal
- simultaneous
- stay put
- temporary simultaneous

If temporary simultaneous is chosen then further information will need to be given when any evacuation strategy becomes permanent.

Fire and smoke control equipment in the residential units of this building or structure

Include all equipment provided in the residential units e.g.

- heat detectors
- smoke detectors
- sprinklers or misters

If there are none of these types of equipment in the residential units, additional information should be included.

Fire and smoke control equipment in the parts of this building or structure that residents share

Include all equipment provided in the parts of the building shared by all residents e.g.

- alarm connected to heat or smoke detectors
- alarm connected to call points
- fire dampers or duct dampers
- fire extinguishers
- fire shutters
- heat detectors
- dry risers
- wet risers
- automatic smoke control systems (also known as AOVs or automatic opening vents)
- manual smoke control systems

- smoke detectors
- water sprinklers or misters.

Again, if no such equipment is provided, information should be given.

Where are each type of fire and smoke equipment in this building or structure?

This question applies for each type of fire or smoke equipment you have. Select all that apply

- basement
- bin store
- car park
- common balcony
- common corridor
- common staircase
- external staircase
- lobby
- shared space with equipment, like a utility room, meter room, or shared laundry room
- shared space with no equipment, like a communal lounge
- rooftop
- other – if you select this option, we will not ask for more details about these other locations.

Types of lift in the building or structure

Select all that apply

- evacuation lift – these have structural, electrical and fire protection. They are used to evacuate people with a disability or extra needs
- firefighters lift – these have protection, controls and signals for firefighters to control. The 2003 British Standard classifies these lifts. So, a lift installed before 2003 will not be a firefighters lift
- firefighting lift – these are like firefighters lifts but the standards are less stringent. The 1986 British Standard classifies these lifts. So, a lift installed before 1986 will not be a firefighting lift
- modernised lift for fire service use – these are normally used for passengers. But they have basic extra protection, with controls and signals for firefighters to use
- firemen's lift – these were installed before firefighting lift standards became available (in 1986). They have no complex lift controls or protection measures for firefighters to use.

If the building has no lifts that assist with firefighting or evacuating residents, this information should be recorded.

Number of residential unit front doors in this building or structure

You should count the residential front doors and state how many

- have no certified fire resistance – use this answer if a residential front door is not a fire door
- are certified fire resistant for 30 minutes
- are certified fire resistant for 60 minutes
- are certified fire resistant for 120 minutes
- and how many of the residential front doors there are where you do not know the level of fire resistance.

Number of fire doors in the residential common parts of this building or structure

Count fire doors that residents can walk through, like those between corridors and staircases.

Information is required for how many of the fire doors are certified as fire-resistant for

- 30 minutes
- 60 minutes
- 120 minutes
- and how many fire doors there are where certified fire resistance is unknown.

Energy supplies, storage and generation

Storage of energy that supplies the building or structure must be given. Energy that supplies individual resident's equipment does not need to be included.

Types of energy storage could be

- hydrogen batteries
- lithium ion batteries
- other – select this if there are other types of energy storage on site.
- no types of energy storage on site.

Onsite energy generation for this building or structure

Types of onsite energy generation could be

- air or ground source heat pumps
- biomass boiler
- solar panels or wind turbines
- other
- none of these types of energy generation.

Energy supplies to the building or structure

Types of energy supply could be

- district or communal heating
- mains electricity supply
- mains gas supply
- oil
- other.

Type of structure
What is the type of structure?

- concrete and steel
- concrete large panel system – 1960s
- concrete large panel system – 1970 onwards
- modular – concrete
- concrete – other
- lightweight metal structure, like aluminium
- masonry
- modular – steel
- steel frame
- modular – other metal
- modular – timber
- timber
- none of the above.

Reinforced concrete
If the structure of the building is reinforced concrete, select

- concrete and steel – if the concrete is reinforced with steel
- modular – concrete – if reinforced concrete is used in a modular structure
- concrete – other – if the concrete is reinforced by any material other than steel.

If the structure of your building is a reinforced concrete large panel system (LPS), select the relevant LPS option for the year it was built

- during the 1960s
- from 1970 onwards.

Roof
Type of roof on the building or structure

- flat roof
- pitched roof
- mix of flat and pitched.

Does the roof structure of this building or structure have a layer of insulation?

- yes, on top of the roof structure
- yes, below the roof structure
- no.

What material covers the largest surface area on this building or structure's roof?

Select all that apply

- composite panels
- fibre cement asbestos
- metal sheet – this includes aluminium, zinc or lead
- plastic sheet – this includes PVC, UPVC or fibre reinforced plastic
- polycarbonate sheet
- other sheet material
- rolled or liquid bitumen felt
- rolled or liquid felt – other than bitumen
- rolled or liquid rubber
- rolled or liquid hot or cold roof systems
- reinforced autoclaved aerated concrete (RAAC) – this is for RAAC only. For any other type of concrete, select 'none of these'
- shingles
- slate
- tiles
- green roof (also known as a living roof or sedum roof)
- none of these materials are used on the roof.

Staircases

Total number of staircases

- count all the staircases on the inside and outside, including those that serve the basement
- count the number of internal staircases that serve all floors from ground level to the top floor.

External walls

What materials are visible on the outside of the walls of this building or structure?

Select all that apply

- aluminium composite materials (ACM)
- high pressure laminate (HPL)

- metal composite panels
- other composite panels
- concrete
- green walls
- masonry
- metal panels
- render
- tiles – any type of tile, including brick slip, ceramic or rubber
- timber
- other.

If 'aluminium composite materials (ACM)' or 'high pressure laminate (HPL)' are selected, provide information on whether the material

- meets the fire classification A2-s1, d0 or better (set out in the British Standard BS EN 13501-1, this classifies building materials based on their reaction to fire)
- has passed a large-scale fire test to BS 8414 (the British Standard for tests that assess the fire safety of cladding applied to the outside of buildings)
- does not meet the fire classification and has not passed the large-scale fire test standard.

Percentage of each material on the outside walls of this building or structure

You should estimate the percentage of each material. They must all add up to 100%.

For example, 70% might be covered in concrete while the remaining 30% might be covered in glass.

What type of insulation is used in the outside walls of this building or structure?

Select all that apply

- fibre insulation – glass or mineral wool
- fibre insulation – wood or sheep wool
- foil bubble or multifoil insulation
- phenolic foam
- polystyrene insulation – expanded polystyrene (EPS) or extruded polystyrene (XPS)
- polyurethane (PUR) or polyisocyanurate (PIR or ISO)
- other – provide information on what type of insulation materials were used but do not use brand names
- none of these.

Estimate the percentages of each type of insulation material. They must all add up to 100%.

Features on this building or structure

Select all that apply

- advertising hoarding attached to a wall
- balconies
- communal recreation area on the roof
- communal walkway between structures
- escape route onto and across the roof
- external staircases
- machinery in an outbuilding
- machinery in a room on the roof
- machinery on the roof
- phone masts
- roof lights or sky lights
- solar shading
- other
- none of these features.

Which materials are used most in the features of this building or structure?

Select no more than two of these materials for each of the features listed in the previous question

- aluminium
- concrete
- glass
- masonry
- metal
- plastic
- slate
- timber
- other.

How is the building used?

The list of options includes

- assembly and recreation – any place of assembly, entertainment or recreation, like a bingo hall, theatre, dance school, gym or public library
- office – any office, bank or building society, police or fire station
- residential dwellings, like flats or maisonettes
- residential institution, like a care, residential or nursing home, prison, hospital or boarding school
- other residential use, like a hotel, boarding house, halls of residence or hostel

- shop and commercial – any retail business, like a shop, restaurant or hairdresser
- other non-residential – any other non-residential use, like a car park, storage, bin or meter room, or factory.

Primary use for this building or structure

Choose one item from the list of potential building uses.

Secondary uses for this building or structure

Select all that apply from the list of potential building uses.

There is also an extra option for you to tell us that the building has no secondary uses.

The primary and secondary uses cannot be the same.

Number of floors below ground level in this building or structure

Count any floor whose ceiling is below ground level, like a basement.

Ground level is the level of land immediately next to this building or structure. If that land is uneven, count from the lowest part of the land immediately next to this building or structure.

Any ceiling below that level is below ground level. If any part of the ceiling is at or above ground level, count the floor as above ground level.

Primary use of the floors below ground level in this building or structure

If the floors below ground level have multiple uses, select the option that takes up the most floor space.

Has this building or structure had a different primary use in the past?

This question applies if the primary use for the building is 'residential dwellings'.

The Building Safety Regulator needs to know if this building previously had a different primary use. For example, an office or factory could have been converted into residential units.

What was the previous primary use of this building or structure?

If it had multiple uses, select the option that took up the most floor space.

What year did this building or structure change use? (optional)

This is the year on the Completion Certificate that was issued when this change of use happened.

Building works since the original build

Provide information, for example from the building's records.

Building works since this building or structure was originally built

Include information on

- asbestos removal and remediation on multiple floors
- balconies added
- changes in number of residential units
- changes in number or location of staircase cores
- changes to windows
- complete rewiring of buildings or floors
- floors added
- floors removed
- reinforcement works to large panel system structure
- work connected to external walls – includes replacement of cavity barriers, insulation or cladding (like ACM).

Select from building works that involved the installation, replacement or removal of

- active or passive fire systems
- emergency or other lighting
- cold water systems in multiple residential units
- heating or hot water systems in multiple residential units
- gas supply to the building.

If you do not know about any building works that needed building control approval or the building has not had any of the building works listed above, the Building Safety Regulator will need confirmation.

Connections

Include information on the way buildings and structures can be connected

- bridge or walkway between sections
- car park below ground
- ground floor, like a shared entrance hall
- levels below ground with a residential unit
- levels below ground without a residential unit
- shared wall with emergency door
- shared wall with everyday use door
- shared wall with no door
- other.

What connects the structures of this building?

Connections between this building and other high-rise residential buildings

A building may be connected to another by a shared wall, basement, lobby or walkway.

Is this building connected to any other high-rise residential buildings?

What connects this building to the other high-rise residential buildings?

Is your building connected to any other buildings?

What connects your building to the other buildings?

Put measures in place to prevent building safety risks happening and reduce the severity of any incident that does happen

Every Accountable Person is under an ongoing obligation to carry out an assessment of the building safety risks in/of/arising from, and relevant to, the part of the higher-risk building for which they are responsible. They must take all reasonable steps and actions necessary to manage the building safety risks, preventing them from materialising in their building and to mitigate the severity of such an incident if it does occur.

Building safety risks

A 'building safety risk' is defined as any risk to the safety of persons in or about a building arising from fire, structural failure or any other matter prescribed in regulations.

The clause is drafted to capture risks inside and outside the building, such as fire spreading on the external wall of a building, but is limited to risks that are associated with the building and/or arising from the building itself. Risks are also limited to those affecting people in and about the building.

The explanatory notes to the Act state that the scope is only intended to encompass major and rapid onset events and that the government 'has no current plans for the higher-risk building regime to extend beyond the risks of fire and structural integrity.'

However, the Secretary of State can request advice from the Building Regulator about new and emerging risks and may introduce additional definitions for building risk.

(Guidance: Building Safety Regulator: Giving us structure and safety information (key building information. Updated 10 October 2023)

What are some of the measures necessary to comply with the duty of taking all reasonable steps and actions necessary to manage the building safety risks?

High-rise residential buildings can be complex and a wide range of experience and technical skills will be needed to fully evaluate building safety risks. Putting together a team of competent professionals to review the information available and to establish what additional information is necessary is a good first step.

Formulate an Action Plan to review the building and its services. A well-used approach is to commence with HAZID – hazard identification. This is often referred to as the 'what if' study. What happens if 'x' happens, who would be affected, how severely, and so forth. Hazards are identified and what happens if the hazards develop is explored.

In the first instance, the Principal Accountable Person/Accountable Person should review all the information that currently exists about their building.

The Health and Safety File prepared in accordance with the Construction (Design and Management) Regulations 2015 (if available) will contain vital information.

The Fire Risk Assessment will provide fire safety information.

O & M manuals, maintenance records, and so forth, will all contain valuable information.

Once hazards have been identified, consider how these may create risks, what level of risk will be created, the consequences of those risks, and so on. Then determine what needs to be done to mitigate the risks, reduce the hazards and protect people before buildings.

All reasonable steps should be taken to prevent and mitigate building safety risks. What is reasonable depends on the individual circumstances of the building. You should consider

- measures already in place and how effectively they control building safety risks
- what other measures could be taken and whether they are reasonable.

Certain factors will influence whether measures are reasonable – for example, whether they may create additional risks or are disproportionately expensive.

Setting up a safety management system for high-rise residential buildings

S.84 of the Building Safety Act 2022 requires Accountable Persons to take steps to manage building safety risks in accordance with prescribed principles by setting up a safety management system. The Higher-Risk Buildings (Management of Safety Risks etc.) (England) Regulations 2023 provide further clarity on such 'prescribed principles'. (NB: See earlier question.)

A Safety Management System is a formal management system or framework for managing safety risks.

One simple model for a Safety Management System is based on the Plan, Do, Check, Act approach, the safety management process advocated by the HSE for general health and safety management. This model covers

- Plan: determine your policy and plan for implementation
- Do: assess risks and implement your plan
- Check: measure performance
- Act: review performance and act on lessons learned.

Keep things simple and proportionate. Focus on the main hazards and risks. Things that might influence you are the

- complexity of your building – is it a mixed-use building, or is it a complex design structurally?
- complexity of the measures in place to control building safety risks – consider how many safety measures you have in place, and how they integrate
- number of buildings you manage
- size of your organisation
- extent to which contractors and third parties manage or maintain the measures in place
- resident profile – for example, do any of your residents need help to evacuate or have any additional needs?

Documents produced should be concise and easily understood.

Introduce procedures for checking and reviewing the hazards and risks identified such as audit programmes, maintenance checks and continuous improvement schemes.

Keep records of findings and actions taken.

Engage with senior management and include the safety management plan review on Board agenda, management team meetings, client liaison meetings, and so forth.

Report certain fire and structural safety issues or incidents.

As an Accountable Person, you must submit mandatory occurrence notices and reports to the Building Safety Regulator for the parts of the high-rise residential building you are responsible for.

Safety occurrence examples when buildings are occupied by residents are as follows

- the spread of fire
- total or partial collapse of the building
- defective building work
- unexpected failure or the degradation of construction materials
- the discovery of structural defects
- failure of a critical fire safety measure, such as an automatic opening vent, smoke extraction or fire doors.

The Principal Accountable Person must establish and operate a mandatory occurrence reporting system for residents and others to report safety occurrences to the Accountable Persons.

Prepare and update the Safety Case Report

The Principal Accountable Person must prepare and update the Safety Case Report.

The Safety Case Report is a document that summarises the Safety Case for a high-rise residential building. It identifies the building's safety risks and explains how the risks are being managed. Building safety risks are the risk of the spread of fire or structural failure.

The Safety Case Report must contain the following.

Accountable Persons – provide information about each Accountable Person, including their

- name
- email address
- telephone number
- postal address in England or Wales.

Responsible Persons – provide the details of the Responsible Persons for the building, under the Fire Safety Order.

Who prepared the report – provide the details of the individuals who prepared the report, such as

- their names
- their job roles and the organisation they work for
- how their skills, knowledge and experience make them suitable for working on the report
- the parts of the report they are responsible for.

Building description – include the building name, full postal address, and a photograph or map of where it is. If the building is part of a wider development, describe it and any shared facilities briefly. Include a description of the nearby area including other buildings and transport routes.

If there is any missing building information, explain what action has been taken to find the missing information.

Risk assessments summary – use the building risk assessments provided by each Accountable Person to put together

- a description of the types of risk assessment techniques used throughout the building
- an overview of who did the assessment and their competence to do so
- the significant findings of the assessments
- a summary of any outstanding actions or recommendations and the plan to progress them.

Managing risks summary

Use the information from each Accountable Person to provide a description of

- the measures in place to prevent and mitigate building safety risks for the whole building
- how the measures work to prevent or mitigate risk
- assessments that have been done which have identified building safety risks and what changes need to be made to mitigate the risks
- conclusions from reports that have been done, for example, a structural inspection.

Safety management system

The Safety Case Report should include a summary of the safety management systems in place for the whole building to manage building safety risks. This is essential to show how risks are being properly managed. To do this, use information from each Accountable Person about the management system they use in their parts of the building.

Planning for emergencies

In the Safety Case Report, describe the strategy for the building during an emergency. Explain why this approach has been chosen and how it works. For example, your building may be designed to support a stay put strategy and it works by relying on the controls in place, such as compartmentation.

If the strategy changes from what was originally intended, you should explain why and describe any additional measures in place to support the change.

Ongoing work and building improvement

While gathering information for the Safety Case Report, you might find control measures that are missing or other issues, like ongoing building work, that affect the building's safety.

Describe the work that needs to be done to resolve these issues, how it will be done and the timescale for completion. Describe any temporary measures that are in place to manage the risks in the meantime and the timescales to complete the work.

Updating the Safety Case Report

You should update your Safety Case Report, for example, when

- improvement work is carried out to manage building safety risks
- work on the building impacts building safety risks.

Notify the Building Safety Regulator of any updates or revisions to your Safety Case Report.

Prepare a residents' engagement strategy setting out how residents will be involved in building safety decisions

The Principal Accountable Person must

- prepare a residents' engagement strategy
- act in accordance with the strategy
- review and revise the strategy and keep a record of the reviews
- provide the latest version to each Accountable Person
- when necessary, consult residents, owners of residential units and Accountable Persons about the strategy and take their opinions into account.

Accountable Persons, for the parts of the building that they are responsible for, must

- provide the latest version of the strategy to residents and owners of residential units
- tell residents about building safety work
- tell them about who will carry out the work.

Accountable Persons must work with other Accountable Persons for the building to help prepare and review the strategy.

Set up and operate a system to investigate relevant complaints relating to building safety risks or the performance of an Accountable Person's duties

The Principal Accountable Person of a high-rise residential building must establish and operate a complaints system for residents, and other users of the building, who want to raise an issue about

- structural failure and spread of fire in the building (building safety risks)
- the performance of an Accountable Person.

The complaints system must enable residents, and others, to raise issues with those responsible for managing building safety risks.

Residents should be able to understand what a relevant complaint is, so that they know how to raise issues when they need to.

The complaints system should help Accountable Persons to treat all issues raised impartially and fairly. Residents and other building users should not be afraid of raising issues.

The complaints policy must be published. It must explain to residents what a 'relevant complaint' is. It should set out the general approach to issues raised and how the complaints system will work. It should explain how the Principal Accountable Person will manage issues raised. It should clearly say who does what, when and how.

A 'relevant complaint' is

- structural failure and spread of fire (building safety risks)
- performance of an Accountable Person.

You must tell residents and others

- what a relevant complaint is
- how to raise issues
- how the issues will be dealt with
- how long it will take to investigate issues
- when they should expect responses at specific stages and what they will cover
- how you will collect and store information in line with General Data Protection Regulation (GDPR).

Describe how and when issues can be taken to the Building Safety Regulator, providing contact details to allow those issues raised to be escalated.

Those raising an issue must be able to challenge responses, and you should tell them how they can do this. When necessary, residents can appoint a representative to raise an issue for them.

Consider consulting with residents on any changes to your system.

Existing complaints procedures may be suitable for amendment and adaptation to reflect the needs of the Building Safety Act 2022.

The complaints procedure system must be shared with all Accountable Persons and made available to residents.

Apply for a Building Assessment Certificate when directed by the Building Safety Regulator to do so

Once a higher-risk building has been registered with the Building Safety Regulator, the Principal Accountable Person will eventually be invited by the Regulator to apply for a Building Assessment Certificate. A Building Assessment Certificate must be obtained for every building registered.

The Principal Accountable Person will have 28 days from the day on which the Building Safety Regulator makes that call to submit its application, and if they fail to do so, they will have committed an offence under s.79(3) with the risk of summary conviction and imprisonment.

There will be a charge for the application and an hourly rate charged by the Building Safety Regulator to check and review the information submitted.

Applications must include

- a copy of the latest Safety Case Report
- a copy of the residents' engagement strategy
- information on the mandatory occurrence reporting system
- a statement confirming that all Accountable Persons have supplied relevant information to all appropriate persons.

The application must include details of the building – that is, address, registration number, name of the Principal Accountable Person, details of any changes to the information held on the register and details of any compliance notices issued in relation to the building.

Under Section 81(3)(b) of the Building Safety Act 2022, the Building Safety Regulator will refuse a Building Assessment Certificate if it is not satisfied that all the relevant duties have been complied with. Alternatively, s.81(4) allows the regulator to issue a notice to the Principal Accountable Person with a brief description of the contravention, specifying a period for remedying the contravention. s.81(4) will only apply to remedies that can be implemented promptly, allowing the Principal Accountable Person to attempt to remedy the issue and still be awarded a Building Assessment Certificate, rather than submitting a new application.

It is not yet clear what happens when an application for a Building Assessment Certificate is refused and how long an applicant will be given to re-submit the application, particularly while the building remains occupied. It is likely that the Building Safety Regulator will use its powers under s.99 of the Building Safety Act 2022, which allows them to submit a 'compliance notice'. A compliance notice will specify the steps a Principal Accountable Person should take over a particular period. The notice will be proportionate to the risk identified, and if the application identifies any immediate danger, then under s.99(4) the Building Safety Regulator will be able to issue an 'urgent action notice'.

Miscellaneous expectations

Accountable Persons must be aware that newly constructed higher-risk buildings or those that have undergone substantial refurbishment must have a Completion Certificate issued by the Building Safety Regulator before the building is occupied.

Accountable Persons and Principal Accountable Persons must either have the relevant competence or appoint someone with the relevant competence to help carry out their duties.

Where Accountable Persons and Principal Accountable Persons appoint others to help carry out their duties, they should continually monitor and assess their competence and capacity.

Individuals must have the appropriate level of skills, knowledge, experience and behaviours to effectively perform their duties. Organisations must have the right organisational capability.

Principal Accountable Persons and Accountable Persons must transfer building safety information to any incoming Accountable Person.

Principal Accountable Persons and Accountable Persons must notify the Building Safety Regulator if there is a change to an Accountable Person.

Display a notice in the building containing information about the Accountable Persons for the building, including their full contact details and a summary of their roles and duties.

Register all new higher-risk buildings before occupation with the Building Safety Regulator and receive a Completion Certificate.

I am an Accountable Person – what golden thread requirements do I need to be aware of if I am responsible for an occupied higher-risk building?

The Building Safety Act 2022

For the golden thread of information, the Act

- specifies that Accountable Persons must keep certain information as golden thread information and must keep this information to the required standards
- specifies that that this information must be shared with certain persons
- also gives the Secretary of State the power to make regulations to make further requirements around what information is kept as golden thread information, the sharing of this information and how it is kept. This applies to any higher-risk building that is occupied. This is set out in sections 88, 89 and 90 of the Act.

The Higher-Risk Buildings (Management of Safety Risks etc.) (England) Regulations 2023

These Regulations place requirements on Accountable Persons, residents and others with regard to building safety and the management of building safety duties for occupied higher-risk buildings.

For the golden thread of information, the Regulations

- set out how the Accountable Persons must keep the golden thread of information during occupation of a higher-risk building. The Regulations set out that the golden thread must be accessible, accurate, accountable, electronic, secure, transferable and understandable (Regulation 7). The Regulations do not specify that the golden thread has to be up to date as this is already specified in the Building Safety Act
- specify detailed requirements around the provision of golden thread information to the Building Safety Regulator, residents and others (Regulations 8, 9 and 11). This includes setting out when the information must be given and the form it must be given in
- set out the prescribed principles for the ongoing management of building safety risks and the content of Safety Case Reports (which will be kept as golden thread information) (Regulations 4 and 5).

The Higher-Risk Buildings (Keeping and Provision of Information etc.) (England) Regulations 2024

These Regulations specify the golden thread information that Accountable Persons for a higher-risk building must keep and the information that they have to share with other persons (including residents).

The Regulations specify

- the golden thread information that Accountable Persons should keep during occupation (Regulation 4 and Schedule 1)
- what information needs to be handed over when an Accountable Person leaves their role (Regulation 13)

- the information that should be shared with other persons and when this information should be shared (Regulations 5–12 and 14). These persons are specified in the Act and these Regulations as
  - the Building Safety Regulator
  - Responsible Persons
  - fire and rescue service
  - residents
  - owners of flats
  - Clients
  - landlords of residential units
- certain exceptions to the sharing of information (Regulations 15–19).

The Building Safety (Registration of Higher-Risk Buildings and Review of Decisions) (England) Regulations 2023

These Regulations set out

- the registration information that a Principal Accountable Person must keep and must provide to the Building Safety Regulator when they register their higher-risk building (Regulations 11–19)
- that when the registration information is updated, the Principal Accountable Person must notify the Building Safety Regulator within 14 days (Regulation 4).

The Higher-Risk Buildings (Keeping and Provision of Information etc.) (England) Regulations 2024 specify that the registration information must be kept as golden thread information.

The Higher-Risk Buildings (Key Building Information etc.) (England) Regulations 2023

These Regulations set out

- the key building information that Principal Accountable Persons must keep (Regulations 3–18) and information they must submit to the Building Safety Regulator within 28 days of registration (Regulations 19 and 20)
- that when the information is updated, the Principal Accountable Person must notify the Building Safety Regulator within 28 days (Regulation 21)
- that the key building information must be kept as golden thread information. This is specified in The Higher-Risk Buildings (Keeping and Provision of Information etc.) (England) Regulations 2024.

The Higher-Risk Buildings (Descriptions and Supplementary Provisions) Regulations 2023

The term 'higher-risk building' in occupation is defined in Part 4, Section 65(1) of the Building Safety Act 2022 as a building in England that is at least 18 m in height or has at least seven storeys.

These Regulations supplement that definition and specify that a higher-risk building in occupation also has to include two residential units.

I am a duty holder – what golden thread requirements do I need to be aware of if there is building work in an existing higher-risk building?

The Building (Higher-Risk Buildings Procedures) (England) Regulations 2023

These Regulations introduce a new building control process for building work in an existing higher-risk building. This work is overseen by the Building Safety Regulator.

The Regulations provide the detail of the building control regime and the requirements when building work is being done to an existing higher-risk building (known as higher-risk building work).

Part 4 of these Regulations specifically sets out the detailed regulations relating to the provision of the golden thread for new higher-risk buildings.

For the golden thread of information, the Regulations specify

- the information and documents that the Client must ensure are kept as part of the golden thread of information before building work begins on a higher-risk building (Regulations 31(3), (4), (6))
- how this golden thread information must be kept (Regulations 31(1) and (2))
- that the Client must ensure that the Principal Designer and Principal Contractor have access to the golden thread information (Regulations 31(1) and (2))
- how and when the information must be updated throughout construction and who is responsible for updating the information (Regulations 31(4) and (6)–(11))
- what information must be handed over on building completion and when this information should be handed over (Regulations 38 and 39)
- how the information should be handed over on building completion and who to (Regulations 38 and 39).

The Building Regulations etc. (Amendment) (England) Regulations 2023

These Regulations amend the Building Regulations 2010 to support the new higher-risk building control regime, and provide for the Building Safety Regulator to be the only building control authority for all higher-risk buildings.

The changes to the building control regime include a new building control process for higher-risk building work, and changes to non-higher-risk building work to raise standards across the built environment. These include clear legal responsibilities for those involved in the commissioning, design and construction of building work, and those who carry out building control.

The Regulations set out requirements with regard to duty holders and competence.

For the golden thread of information, the Regulations specify

- the Client must make suitable arrangements for planning, managing and monitoring a project so as to ensure compliance with all relevant requirements (including the golden thread) (Regulation 11 A)

- the Client must provide building information as soon as is practicable to every designer and contractor on the project (Regulation 11B)
- the Client must ensure designer and contractors are aware that when a project includes higher-risk building work, relevant requirements apply (Regulation 11 B)
- the Client must appoint a Principal Designer and a Principal Contractor and must take all reasonable steps to ensure that the person appointed to the role of Principal Designer and Principal Contractor meets the competence requirements for that role (Regulations 11 D–I)
- the duties of duty holders (Regulations 11 J–O).

The Building Safety (Registration of Higher-Risk Buildings and Review of Decisions) (England) Regulations 2023

These Regulations set out

- that when the registration information is updated, the Principal Accountable Person must notify the Building Safety Regulator within 14 days. Certain significant building work may require registration information to be updated.

The Higher-Risk Buildings (Keeping and Provision of Information etc.) (England) Regulations 2024 specify that the registration information must be kept as golden thread information.

The Higher-Risk Buildings (Key Building Information etc.) (England) Regulations 2023

The Regulations set out

- that when the key building information is updated, the Principal Accountable Person must notify the Building Safety Regulator within 28 days. Building work to an existing building may mean that the key building information needs to be updated and submitted to the Building Safety Regulator

- that the key building information must be kept as golden thread information. This is specified in The Higher-Risk Buildings (Keeping and Provision of Information etc.) (England) Regulations 2024.

The Higher-Risk Buildings (Management of Safety Risks etc.) (England) Regulations 2023

These Regulations place requirements on Accountable Persons, residents and others with regard to building safety and the management of building safety duties for occupied higher-risk buildings. These duties continue even when there is building work in the building.

For the golden thread of information, the Regulations

- set out how the Accountable Persons must keep the golden thread of information during occupation of a higher-risk building. The Regulations set out that the golden thread must be accessible, accurate, accountable, electronically secure, transferable and

understandable (Regulation 7). The Regulations do not specify that the golden thread has to be up to date as this is already specified in the Building Safety Act
- specify detailed requirements around the provision of golden thread information to the Building Safety Regulator, residents and others including any Client for building work (Regulations 8, 9 and 11). This includes setting out when the information must be given and the form it must be given in
- set out the prescribed principles for the ongoing management of building safety risks and the content of Safety Case Reports (which will be kept as golden thread information) (Regulations 4 and 5). The management of building safety risks could change when there is building work ongoing in the building.

The Higher-Risk Buildings (Keeping and Provision of Information etc.) (England) Regulations 2024

These Regulations specify the golden thread information that Accountable Persons for a higher-risk building must keep and the information that they have to share with other persons (including the Client for any building work).
The Regulations specify

- the golden thread information that Accountable Persons should keep during occupation (Regulation 4 and Schedule 1)
- what information needs to be handed over when an Accountable Person leaves their role (Regulation 13)
- the information that should be shared with other persons and when this information should be shared (Regulations 5–12 and 14). These persons are specified in the Act and these Regulations as the Building Safety Regulator, residents, owners of flats, clients, Responsible Persons, fire and rescue service, and landlords of residential units
- certain exceptions to the sharing of information (Regulations 15–19).

The Higher-Risk Buildings (Descriptions and Supplementary Provisions) Regulations 2023

The term 'higher-risk building' is defined in Part 4, Section 65(1) of the Building Safety Act 2022 as a building in England that is at least 18 m in height, or has at least seven storeys.

These Regulations supplement that definition and specify that higher-risk buildings in occupation also have to include two residential units.

Legislation

Building Safety Act 2022

Secondary legislation for higher-risk buildings

- The Building (Higher-Risk Buildings Procedures) (England) Regulations 2023.
- The Higher-Risk Buildings (Descriptions and Supplementary Provisions) Regulations 2023.
- The Building Safety (Registration of Higher-Risk Buildings and Review of Decisions) (England) Regulations 2023.

- The Higher-Risk Buildings (Key Building Information etc.) (England) Regulations 2023
- The Building (Public Bodies and Higher-Risk Building Work) (England) Regulations 2023.
- The Building (Higher-Risk Buildings Procedures) (England) Regulations 2023
- The Higher-Risk Buildings (Management of Safety Risks etc.) (England) Regulations 2023.
- The Higher-Risk Buildings (Keeping and Provision of Information etc.) (England) Regulations 2024.

Guidance documents from the Department for Levelling up, Housing and Communities

- The role and responsibilities of Accountable Persons and Principal Accountable Persons – Building safety campaign, UK Government.
- Safety in high-rise residential buildings: Accountable Persons.
- Assessing safety risks in high-rise residential buildings: a detailed guide.
- Managing safety risks in high-rise residential buildings: a detailed guide.
- Safety management systems for high-rise residential buildings.
- Applying to register a high-rise residential building.
- Operating a complaints system for building safety in a high-rise residential building.
- Operating a mandatory occurrence reporting system.
- Register a high-rise residential building.
- Complain about a building safety risk in a high-rise building.
- Safety case for a high-rise residential building.
- Preparing a Safety Case Report.

The Building Safety Act 2022: Questions and Answers: A practical approach to design and management of higher-risk buildings

Pat Perry
ISBN 978-1-83662-467-7
https://doi.org/10.1108/978-1-83662-464-620251011

Chapter 9
The Principal Designer

The Building Safety Act 2022 has introduced new duty holder roles. I understand that one of these duty holders is to be called the Principal Designer. What does this mean?

As part of the reforms to building control procedures and to the strengthening of the regulatory regime in respect of higher-risk buildings, the Building Safety Act 2022 and its subsidiary legislation has introduced, as of 1 October 2023, a new duty holder regime that applies to all construction projects except for a few limited exceptions.

New duties are proposed for those who procure, plan, design, manage and undertake building work. These apply to all building work to which the Building Regulations 2010 (now amended by the Building Regulations etc. (Amendment) (England) Regulations 2023) apply – not just higher-risk buildings. This is to ensure a stronger focus on compliance with regulations for design and building work, with additional obligations for higher-risk buildings.

The new duty holders introduced under the Building Regulations etc. (Amendment) (England) Regulations 2023 will be the Client (including domestic client), the Principal Designer and the Principal Contractor as well as duties on designers (including the Sole or Lead Designer) and contractors (including Sole Contractor).

This means that for the majority of construction projects, a 'Principal Designer' and 'Principal Contractor' must be appointed, each carrying their own specific duties and competency requirements.

Will the Principal Designer appointed under the Building Regulations be the same as the Principal Designer appointed under the Construction (Design and Management) Regulations 2015?

No, not necessarily. The two different roles could be undertaken by the same organisation or individual or they could be different appointees.

The legal duties of both duty holders are different and they may not always be interchangeable.

It is unlikely that someone appointed under the Construction (Design and Management) (CDM) Regulations 2015 with expert health and safety experience will have the competency in design work necessary under the Building Regulations unless they have architectural, structural engineering, civil engineering or building surveying qualifications.

Why is there a need for two different duty holders both called the Principal Designer?

Dame Judith Hackitt's inquiry into the Building Regulations and fire safety carried out post the Grenfell Tower fire revealed a series of systemic failures in design, construction, certification, management and maintenance processes, triggering widespread regulatory reforms aimed at preventing such tragedies in the future. She found that 'the roles and responsibilities of those procuring, designing, constructing and maintaining buildings are unclear'.

The role of the Principal Designer was initially introduced with the CDM Regulations 2015, which sought to ensure that health and safety considerations were embedded in the design process.

In response to the success of CDM Regulations 2015, Dame Judith Hackitt recommended further strengthening the role of duty holders, including clients, designers and contractors. As part of this update, the Building Safety Act 2022 introduced a second, distinct Principal Designer role focused specifically on Building Regulations compliance.

What is the role of the Principal Designer as required by the Building Safety Act 2022?

The Building Safety Act 2022 introduced secondary legislation, the Building Regulations etc. (Amendment) (England) Regulations 2023 that, as expected, amended the Building Regulations 2010.

A new Part 2A has been introduced to the 2010 Regulations relating to duty holders and competence.

A new section 11D has been inserted into the original Regulations stating

> 11D(1) Where there is more than one contractor, or it is reasonably foreseeable that more than one contractor will be working on a project, the client must appoint in writing—
>
> (a) a designer with control over the design work as the principal designer for the purposes of these Regulations, and
> (b) a contractor with control over the building work as the principal contractor for the purposes of these Regulations.

Principal Designers must

- plan, manage and monitor design work, and cooperate, coordinate and communicate to ensure the design work, if built, complies with the Building Regulations
- understand the law (role and competency requirements set out in the Building Regulations) as this will enable the Principal Designer to understand their role
- be able to manage other designers and reach consensus that design work is compliant with the Building Regulations

- monitor identified compliance risks and assess gaps in other designers' competences
- set the plan for achieving design work compliance
- control changes to the plan and monitor risks to design work compliance
- coordinate designers' work related to achieving the plan
- liaise with Principal Contractors about design work compliance
- monitor progress against the plan through to the end of the design phase
- understand technical guidance, codes of practice and standards so that they can assess, challenge and build design team consensus on design compliance.

The Principal Designer must also

- liaise with the Principal Contractor and share with the Principal Contractor any information relevant to the planning, management and monitoring of the building work, and the coordination of building work and design work for the purpose of ensuring compliance with all relevant requirements
- where the Principal Contractor provides comments to the Principal Designer in relation to compliance with the relevant requirements, have regard to those comments
- if requested, assist the Client in providing information to other designers and contractors.

The Principal Designer role is concerned with making sure that the finished building is safe and fit for purpose and complies with all relevant sections of the Building Regulations.

Can anyone take on the role of Principal Designer under the Building Regulations?

No.

A new section, 11G, in the Building Regulations states

> (1) A principal designer must have—
>
> (a) where the person is an individual, the skills, knowledge, experience and behaviours necessary,
>
> (b) where the person is not an individual, the organisational capability to fulfil the duties of a principal designer under these Regulations in relation to the design work included in the project.
>
> (2) Where the principal designer (D) is not an individual, D must designate an individual who has the task of managing its functions as the principal designer.
>
> (3) Before making the designation under paragraph (2), D must take all reasonable steps to satisfy themself that the individual to be designated has the skills, knowledge, experience and behaviours necessary to manage the function of principal designer on behalf of D in such a way as to ensure D fulfils the duties of the principal designer under these Regulations in relation to the design work included in the project.

Therefore, the Principal Designer appointed under the Building Regulations must be competent to undertake the duties allocated to the role.

The issue of competency of various professional disciplines was of great concern during the Public Inquiry into the Grenfell Tower fire and several recommendations were made to improve the competence and consistency of approach of members of a project team.

Skills, knowledge, experience and behaviours were not just to be assumed in respect of duty holders – they were to be demonstrated via competency registers, professional body memberships, qualifications, and so forth.

What is the difference between the Principal Designer appointed under the Construction (Design and Management) Regulations 2015 and The Building Regulations etc. (Amendment) (England) Regulations 2023?

The CDM Principal Designer is primarily concerned with the safety of workers and others during the construction process. Their work revolves around identifying and managing health and safety risks and ensuring that everyone involved in the project is aware of their responsibilities regarding safety. Their ultimate goal is to prevent accidents and injuries during construction and to create a building that can be safely maintained in the future. A CDM Principal Designer should raise any issues with designs that create future safety risks for building users or for those who clean and maintain the building. The Principal Designer should challenge other designers on design issues such as changes in floor level, access to plant and equipment, high level cleaning, floor surface slipperiness, risk of legionella from water storage facilities and so on.

The Building Regulations Principal Designer is primarily concerned with the technical compliance of the building itself. Their role ensures that the building will meet all statutory requirements, such as fire safety, structural stability and energy efficiency, once it is completed. Their primary concern is the long-term safety and functionality of the building itself, rather than the safety of the construction process.

When does the requirement to appoint a Building Regulations Principal Designer apply?

The requirement for a Principal Designer under the Building Regulations is set out in amended Building Regulation 11D and applies to all building projects where there is more than one contractor or it is reasonably foreseeable that more than one contractor will be working on the project.

The appointment must be in writing.

It should be noted that if the Client fails to appoint a Principal Designer or Contractor, the responsibilities will be retained by the Client and will default to them.

Where the building works involve a higher-risk building – that is, one that has seven or more storeys and is over 18 m high – the Principal Designer must be named.

The Client – that is, the person commissioning the building project – must appoint the Principal Designer as follows

Regulation 11(D) subsection 3

(a) in relation to a project which includes higher-risk building work for which an application for building control approval must be submitted to the regulator, before that application is submitted;
(b) in relation to any other project, before the construction phase begins.

Can the Principal Designer be changed during the course of a building project?

Yes.

Regulation 11(D) subsection 4 states

> Where the appointment of a principal designer or the principal contractor ends before the end of the project, as soon as reasonably practicable the client must appoint a new principal designer or new principal contractor, as the case may be, under this regulation.

Where a replacement principal designer is appointed, they must review the arrangements the previous principal designer put in place for fulfilling the duties, so that all reasonable steps are taken to ensure that the design is such that if the building work to which the design relates were built in accordance with that design, the building work would be in compliance with all relevant requirements.

What is the definition of Principal Designer?

The amended Building Regulations have a similar definition for Principal Designer as the CDM Regulations 2015 and the Principal Designer is defined as

> a designer with control over the design work (which applies to all buildings in England, requiring building control approval). (Regulation 11D(1))

The CDM Regulations 2015 define the Principal Designer as

> a designer with control over the pre-construction phase.

Under the CDM Regulations 2015, the Principal Designer has overall responsibility for coordinating the health and safety aspects of the design and planning phase and for the pre-construction information pack and for producing the health and safety file.

Who appoints the Principal Designer under the Building Regulations and when?

The Client – that is, the person who commissions the building project.

A Principal Designer must be appointed in writing before the construction phase begins, or before submitting an application for building control approval for a higher-risk building to the Building Safety Regulator.

What additional duties does the Principal Designer have in relation to higher-risk buildings?

There are additional duties if the project involves higher-risk building work, as follows

- establish and operate a mandatory occurrence reporting system
- submit mandatory occurrence notices and reports to the Building Safety Regulator
- store the required information about the building (also known as the golden thread), and provide it to the Client and the Principal Contractor as necessary
- keep and share with the Client the steps you have taken to make sure those you appoint to do design work meet the competence requirements.

How should the Principal Designer demonstrate that they have taken all the steps necessary to ensure that other designers have complied with their duties under the Building Regulations?

A design management system will be important. There is no specific format required by the Regulations and each individual or organisation should develop a system that suits their own project requirements.

A suitable approach may be to initiate various workshops during the project such as

- building regulations workshops
- design coordination workshops
- design risk assessment workshops
- specialist services workshops
- Client progress meetings.

It will be good practice to liaise with the Client about setting up systems and procedures as the Client has the ultimate responsibility to ensure that all the duty holders they appoint fulfil their statutory duties.

The Client is responsible for making appropriate project planning, management and monitoring arrangements, including allocating enough time and resources for compliance with the Building Regulations.

It is the Client's responsibility to

- ensure design work is carried out so that, if built, the building work would be compliant with the Building Regulations
- ensure the building work is carried out in accordance with the Building Regulations
- enable designers and contractors to cooperate to ensure compliance with the Building Regulations
- undertake periodic reviews to identify any higher-risk buildings
- provide building information to every designer and contractor on the project
- ensure arrangements are maintained and reviewed throughout the project
- assess the competency of the Principal Designer and Principal Contractor.

In order to ensure that everyone meets their statutory duties, a collaborative and coordinated approach to project management should be adopted.

There are many similarities with the CDM Regulations 2015 in respect of the Principal Designer's role of coordination designs from a health and safety perspective and ensuring the design risk assessments are completed for project plans, and so forth.

A Building Regulations and Design Risk Register could be designed and updates recorded at each design team/progress/site meeting.

Remember too, that there is a statutory duty to keep an up-to-date Change Control Plan in place for the project and similar information might be transferable to each management system (see Chapter 17 on change control).

Is the Principal Designer under the Building Regulations the same as the Lead Designer on a project?

The two roles can be performed by the same person/organisation or each role could be performed by a different person.

The Royal Institute of British Architects advises that both roles would be best carried out by the same individual or organisation, as this would reduce duplication of roles and cut down on confusion between responsibilities of duty holders and give the Client clear accountability for legal compliance.

The Principal Designer is responsible for ensuring that all applicable regulations are complied with during the course of the project.

Principal Designers must plan, manage and monitor the design work during the design stage of a project, and coordinate matters relating to design work so that all reasonable steps are taken to ensure that the design of the building work would comply with all relevant requirements. This includes ensuring that the other designers of the project are cooperating, coordinating and complying with the relevant requirements.

The Lead Designer is usually responsible for coordinating the overall design and often brings the flair and creativity to a project. The Lead Designer may have specialist knowledge on a critical core element of the design – for example, structural glazing – and may not necessarily be familiar with the detail and legal requirements of other aspects of the design.

The Lead Designer can sometimes be referred to as the Design Coordinator and directs and coordinates other designers in the consultant team as well as any specialist designers that are appointed. This role might include

- coordinating site surveys
- coordinating the preparation of information for the project brief
- coordinating the preparation of designs and specifications
- integrating different aspects of the design and their interfaces into the overall design

- coordinating internal and external consultations and design reviews
- defining the form and content of design information to be prepared
- reporting to the Client on design matters and seeking approvals
- coordinating the preparation of schedules of inspections, tests, mock ups and samples
- coordinating consultations, negotiations and submissions to planning authorities and other statutory and non-statutory authorities
- coordinating the preparation of tender documentation and reviewing submissions
- coordinating quality control systems
- coordinating the issue of production information to contractors and the review of designs prepared by contractors
- coordinating procedures for inspections, commissioning, testing and Client training.

The Principal Designer must take 'all reasonable steps' to ensure compliance with all relevant requirements. What does this mean?

'All reasonable steps' is a new term that is not defined in the Act or accompanying Regulations. It is similar, but not identical, to concepts like 'as far as reasonably practicable' from health and safety law.

One of the challenges is what 'all reasonable steps' looks like and the fact that it will differ from building to building. It is therefore not possible to be prescriptive about what it means in any particular building and this accounts for the fact that there is no specific legal definition within the Building Safety Act 2022 nor in the Building Regulations etc. (Amendment) Regulations 2023.

The Principal Designer will need to demonstrate that they have procedures and processes in place to ensure that they can check if designs comply with all relevant requirements.

First of all, a list of the applicable legislative requirements will be needed – that is, which Building Regulations are applicable to the project, what additional requirements are required if the building is higher risk, what health and safety standards need to be met, which environmental laws are applicable, and so forth.

The Principal Designer will then need to show that they have been able to question and challenge other designers to ensure that they have met their responsibilities and that, as Principal Designer, they are satisfied that there is sufficient evidence available.

Records are important as they will provide the evidence needed for the Regulator or the enforcing authority. Written records or digital records can be presented as evidence, notes on drawings may be sufficient to show that reasonable steps were being taken to check designs and specifications.

The Building Safety Act 2022 section 84(1) and its secondary legislation requires that the Accountable Person takes all reasonable steps to prevent and mitigate building safety risks. The Building Safety Regulator has provided guidance on the term as it applies to the creation of the Safety Case and readers are referred to Chapter 8 on 'Accountable Persons' and Chapter 13 on 'The Safety Case' for further details.

The Principal Designer must liaise with the Principal Contractor and share information with them. What does this mean in practice?

The underlying objective of the Building Safety Act is to improve the flow and quality of information between parties involved in building works and to improve compliance with Regulations and other relevant legislation.

The Principal Designer will not be able to demonstrate that they have taken all reasonable steps to ensure compliance with all relevant requirements if they ignore the Principal Contractor and fail to consult with them about what has been carried out on site and what changes, if any, have been made to designs or materials specifications.

The Principal Designer will need to establish formal processes to review construction activity with the Principal Contractor and to update the Principal Contractor with design changes and other relevant information.

The Principal Designer has a duty to share information relevant to

- the planning, management and monitoring of the building work
- the coordination of building work and design work for ensuring compliance with all relevant requirements of the Building Regulations.

The duty could be exercised by attendance at regular site meetings and the keeping of visit records and meeting minutes.

The Client has requested that I combine the Principal Designer (Building Regulations) and the Principal Designer (CDM) into one appointment. How is it best to do this?

In its document 'Building Safety Competence Information for Principal Contractors and Principal Designers', published in October 2022, the Health and Safety Executive (in its capacity of the acting Building Safety Regulator) states

> A principal designer and principal contractor appointed under CDM 2015 can also be appointed as the principal designer and principal contractor roles for building regulations if competent to do so.

Firstly, you must be satisfied that you have the competencies to undertake both roles as they do not require the same level of knowledge, skills and experience.

If you undertake either or both roles without the requisite competencies, you will be committing an offence and may face fines, imprisonment and reputational damage.

The CDM Principal Designer is responsible for planning, managing and monitoring the pre-construction phase and coordinating matters relating to health and safety during the pre-construction phase. This may involve working with different designers on a project to eliminate foreseeable health and safety risks during construction or taking steps to reduce and control potential risks. The CDM Principal Designer must also consider health and safety risks during the future use of the building and during maintenance and cleaning regimes.

The Building Regulations Principal Designer is required to plan, manage and monitor all design work during the design stage of a project to ensure that the design is compliant with the regulations. The HSE expects these duties to be taken on by the Lead Designer, who must coordinate with other designers to achieve consensus that all design elements are compliant.

Any duty holders appointed by the Client must be confirmed in writing.

Should a Client nominate the CDM Principal Designer as their Building Regulations Principal Designer they must

> certify, in writing, that the person who is the CDM principal designer is treated as appointed as the principal designer for the purposes of the Building Regulations (Regulation 11D(2)), provided the competency requirements are met.

Roles and responsibilities should be clearly identified and, where any additional knowledge and skills are required, they should be clearly set out.

Principal Designers must confirm their appointments in writing and set out their competency and roles and responsibilities for the Client.

It would be wise to set out how you propose to undertake the roles – for example, what type of meetings or project management processes will be implemented to show that coordination, monitoring, review and change control procedures will take place during the project design and construction phases.

A comprehensive 'client care' letter issued to the Client before the formal appointment is made would be good practice as the Client will be clearly advised of the duties and role of the Principal Designer.

It will be important to list out any duties that you will not be undertaking – for example, you may need to advise the Client that you will be seeking advice on all fire safety matters from a competent fire engineer and that this will incur additional costs for the Client. Set out why you believe you do not have the competencies to fulfil this part of the brief.

The Client wishes to replace the initial Principal Designer with a new organisation. Is this permissible and if so, what steps should be taken to comply with the law?

Yes, the Client can make changes to any of the duty holders that they have appointed.

When the Principal Designer's appointment ends, no later than 28 days after the end of the appointment, the outgoing Principal Designer must give the Client a document explaining the arrangements that have been put in place to fulfil the duties of the role.

Where a replacement Principal Designer is appointed, they must review the arrangements the previous Principal Designer put in place for fulfilling the duties, so that all reasonable steps are taken to ensure that the design is such that if the building work to which the design relates were built in accordance with that design, the building work would be in compliance with all relevant requirements.

Where the change in appointment is made after Building Regulations approval or a building notice is given, the Client (or new client) must give a notice to the Building Control Authority in accordance with Regulation 11(O). The notice must be given to the relevant authority within the period of 14 calendar days beginning with the date of the appointment or the date when the person became the Client.

Example letter of appointment for Principal Designer

Dear *********

Statement Of Appointment as Principal Designer under the Construction (Design and Management) Regulations 2015 and the Building Regulations etc. (Amendment) (England) Regulations 2023

Name of project:

In accordance with the legal duties placed on us by the above legislation we, as client on the above project, do hereby appoint [PRINCIPAL DESIGNER'S NAME], of [PRINCIPAL DESIGNER'S ADDRESS], to the position of Principal Designer as required by the Construction (Design and Management) Regulations 2015.

In addition, we also appoint [PRINCIPAL DESIGNER'S NAME] to the position of Principal Designer as required by the Building Regulations etc. (Amendment) (England) Regulations 2023.

The Principal Designer will supply the services as listed in the Specification of Duties attached to this letter of appointment.

The position of Principal Designer in respect of both positions is assumed until the final handover of a project from the Principal Contractor to [CLIENT].

In making this appointment, we, [CLIENT], confirm that we are fully aware of the duties which must be carried out by the Principal Designer and are satisfied that you have demonstrated that you, and/or your company, meet the competency standards required to undertake the roles.

Please sign and return to us the attached Schedule of Services before commencing any duties in respect of the Principal Designer roles.

Signed:

Position:

Date:

Schedule of services

(1) **CDM Regulations: Principal Designer**

As the Principal Designer you will have an important role in influencing how risks to health and safety are managed throughout a project. Design decisions made during the pre-construction phase have a significant influence in ensuring the project is delivered in a way that secures the health and safety of everyone affected by the work.

As Principal Designer you must do the following.

- Plan, manage, monitor and coordinate health and safety in the pre-construction phase.
- Produce and issue to the design team and contractors the pre-construction information pack/report.
- Take in to account relevant information (such as an existing health and safety file) that might affect design work carried out both before and after the construction phase has started.
- Help and advise all parties in bringing together the pre-construction information and provide the information to other designers and contractors as needed in order for them to carry out their duties.
- Work with any other designers on the project to eliminate foreseeable health and safety risks to anyone affected by the work and, where that is not possible, take steps to reduce or control those risks.
- Ensure that suitable design risk assessment team meetings are carried out during the project.
- Establish and manage the design risk assessment procedures and issue suitable templates where necessary or agree with other designers the type and form of design risk assessments.
- Ensure that everyone involved in the pre-construction phase communicates and cooperates, coordinating their work wherever required.
- Liaise with the Principal Contractor, keeping them informed of any risks that need to be controlled during the construction phase.
- Provide expert advice to members of the project team as necessary.
- Assist with checking competency standards of designers and contractors.
- Develop procedures which incorporate a 'Permit to Proceed' for the project so that there is a check to ensure compliance with the CDM Regulations before the project commences on site.
- Provide additional services as follows:

(2) **Building Regulations: Principal Designer**

As Principal Designer you have a legal duty for ensuring that the design of a building complies with the relevant building regulations.

As Principal Designer you must do the following.

- Plan, manage and monitor the design work during the design phase.
- Coordinate all matters relating to design work to ensure that it will comply with the Building Regulations if building work is carried out.
- Ensure duty holder cooperation.
- Ensure all designers comply with their duties.
- Liaise with the Principal Contractor.
- Have regard to comments from the Principal Contractor about compliance with the Building Regulations.
- Assist the Client in providing information to designers.
- Coordinate with the Principal Designer CDM.
- Review the arrangements of any previous Principal Designer.
- Where applicable, sign declarations in respect of the Client's duties when work is complete.
- Perform other duties as follows:

Signed:

Position:

Date:

Checklist for the role of Principal Designer

Building Regulations

Name of Project:			
Address of Project:			
Client Name & Address:			
Design Team:	Discipline:	Date Appointed:	Key contact name:
Who is the Principal Designer?	Name:		
	Organisation:		
When were they appointed?			
Are they part of the design team with direct influence over design decisions?			☐ Yes ☐ No
Has the Client issued confirmation of position in writing?			☐ Yes ☐ No
Have competency assessments for the role of Principal Designer been carried out?			☐ Yes ☐ No
Have design team coordination meetings been set up?			☐ Yes ☐ No
Have Building Regulations coordination or compliance meetings been set up?			☐ Yes ☐ No ☐ N/A
Have communication channels been set up between critical roles?	Client		☐ Yes ☐ No
	Building Control		☐ Yes ☐ No
	Fire Engineers		☐ Yes ☐ No
	Others (Name)		☐ Yes ☐ No
Does the Project involve building or refurbishing a higher-risk building?			☐ Yes ☐ No
Has the Project been submitted to the Building Safety Regulator?			☐ Yes ☐ No

Have the additional duties for higher-risk buildings been identified?	Golden Thread	☐ Yes ☐ No
	Safety Case	☐ Yes ☐ No
	Change Control	☐ Yes ☐ No
	Dangerous Occurrence Reporting	☐ Yes ☐ No
	Residents Engagement	☐ Yes ☐ No
	Fire Safety controls	☐ Yes ☐ No
Name of Principal Contractor:		
Name of individual contact:		
How will information be shared? (please describe)		
How will information be recorded and collated? (please describe)		
How will comments or feedback relating to Building Regulations compliance be actioned? (please describe)		
Has a Change Control Management Plan been implemented?		☐ Yes ☐ No ☐ N/A
Have Design Risk Assessments been provided by all Designers as required by CDM Regulations?		☐ Yes ☐ No ☐ N/A
Is there coordination with the Principal Designer appointed under the CDM Regulations?		☐ Yes ☐ No ☐ N/A
Are procedures in place to ensure compliance with duties relating to the completion of buildings?		☐ Yes ☐ No ☐ N/A

List the information that will be available at the end of the Project:		
Who will the information be distributed to?	Client	☐ Yes ☐ No
	Building Occupier	☐ Yes ☐ No
	Accountable Person	☐ Yes ☐ No ☐ N/A
	Relevant Authorities	☐ Yes ☐ No ☐ N/A
	Building Safety Regulator	☐ Yes ☐ No ☐ N/A
Are you confident that your duties as a Principal Designer (Building Regulations) have been carried out competently?		☐ Yes ☐ No
Project completion date:		
Signed:		
Date:		

The Building Safety Act 2022: Questions and Answers: A practical approach to design and management of higher-risk buildings

Pat Perry
ISBN 978-1-83662-467-7
https://doi.org/10.1108/978-1-83662-464-620251012

Chapter 10
The Principal Contractor

Who or what is the Principal Contractor on a building project?

The Principal Contractor is a contractor appointed by the Client to work on a building project and to take the lead in managing the construction once the project progresses to site.

The Principal Contractor is usually the main contractor and is appointed under Regulation 11D Building Regulations etc. (Amendment) (England) Regulations 2023 (Principal Designer and Principal Contractor) to perform the duties of a Principal Contractor where there is more than one contractor.

What does the Principal Contractor have to do once they have been appointed?

As Principal Contractor, you must do the following.

- Plan, manage and monitor the building work during the construction phase.
- Coordinate matters relating to building work to ensure the building work complies with the Building Regulations.
- Make sure that there is cooperation among all duty holders.
- Make sure all building work is coordinated so that it complies with the Building Regulations.
- Make sure contractors comply with their duties.
- Liaise with the Principal Designer as required.
- Have regard to comments from the Principal Designer concerning compliance with the Building Regulations.
- Assist the Client in providing information to contractors and others.
- Review the arrangements of any previous Principal Contractor.
- Notify building control in writing, where applicable, that the work is being carried out on behalf of a domestic client.
- Sign the declarations as referred to above in the Client's duties when work is complete.

What other duties does the Principal Contractor have?

In addition to the duties listed above, a Principal Contractor must also fulfil the duties of a contractor appointed to a building project.

These include the following.

- Plan, manage and monitor design work so that if the building work were carried out, it would comply with the Building Regulations.

- Cooperate with the Client, designers and contractors to the extent that if the building work were carried out, it would comply with the Building Regulations.
- Not start any design work unless you are satisfied that the Client is aware of the duties owed by the Client.
- Carry out design to ensure that if building work were carried out, the design would comply with the Building Regulations.
- Provide sufficient information about the building's design, construction and maintenance to allow the Client, other designers and contractors to comply with the Building Regulations.
- Consider any other design work and report any compliance concerns relating to the design compliance to the Principal Designer and Client.
- Advise the Principal Designer or the Client whether any work they are designing is higher-risk building work.

Can the Principal Contractor delegate all or some of these responsibilities?

A Principal Contractor is permitted to delegate aspects of its role to third-party consultants and/or subcontractors but it should be noted that legal responsibility for the performance of the Principal Contractor role will remain with the Principal Contractor appointed by the Client, irrespective of such arrangements and the Principal Contractor cannot avoid liability for any failure to deliver the functions of the role by delegating performance to third parties.

What are the differences between the Principal Contractor appointed under the Construction (Design and Management) (CDM) Regulations 2015 and the Principal Contractor appointed under the Building Regulations etc. (Amendment) (England) Regulations 2023?

Under CDM Regulations 2015

The Principal Contractor is primarily responsible for managing health and safety during the construction phase. Appointed by the Client for projects with multiple contractors, their role is to plan, manage and coordinate the construction work, ensuring safety protocols are followed.

Under the Building Safety Act and amended Building Regulations

The Principal Contractor's role extends beyond just the construction phase, emphasising the safety of buildings throughout their life cycle. While they still manage and monitor construction activities, their responsibilities also encompass ensuring the building work complies with the Building Regulations and liaising with designers on any changes.

Is it possible for the Principal Contractor appointed under Construction (Design and Management) (CDM) Regulations 2015 to also be appointed for Building Regulations roles?

Yes, a Principal Contractor appointed under CDM Regulations 2015 can also be appointed as the Principal Contractor role for Building Regulations, provided they have the necessary competence to fulfil those both of roles.

The Client must confirm in writing that both roles are to be undertaken by the same organisation and that they are satisfied that they meet the competency requirements.

Can the same individual or organisation serve as both the Client and the Principal Contractor?

Yes. The same individual or organisation can serve as both the Client and the Principal Contractor, provided they possess the appropriate competencies to undertake the roles.

What are the duties of the Principal Contractor appointed under Construction (Design and Management) (CDM) Regulations 2015?

The Principal Contractor has specific duties under the CDM Regulations 2015 and, ultimately, carries responsibility for site safety issues.

The main duties can be summed up as follows.

- Develop and implement the Construction Phase Health and Safety Plan.
- Appoint only competent and properly resourced contractors to the project – for example, specialist contractors and sub-contractors.
- Obtain and check method statements from contractors.
- Ensure the cooperation and coordination of contractors while they are on site – that is, control multi-occupied site working.
- Ensure health and safety training is carried out.
- Develop appropriate communication arrangements between contractors in respect of site health and safety issues.
- Make arrangements for discussing health and safety issues relative to the project.
- Allow only authorised persons onto the site.
- Display F10 on site for all operatives to be able to see the details.
- Monitor health and safety performance on site.
- Pass information to the Principal Designer for the Health and Safety File.

The Principal Contractor is an important appointment to the construction project as the standard of site safety will be determined by the commitment and competency of the appointment.

What additional duties does the Principal Contractor have in relation to higher-risk buildings?

Where the work is being applied to higher-risk buildings, there are additional duties as follows.

- Establish and maintain a mandatory occurrence reporting system for safety occurrences.
- Collate the relevant golden thread information throughout the construction phase.
- Provide a compliance declaration to the Building Safety Regulator within the Completion Certificate application to confirm you have fulfilled your Principal Contractor functions.
- Keep and share with the Client the steps you have taken to make sure those you appoint to do design work meet the competence requirements.

We have been appointed as a Principal Contractor. How should we approach the duty of coordinating matters relating to building work to ensure the building work complies with the Building Regulations?

As a competent Principal Contractor, you should have systems in place for planning, managing and monitoring works on site. You cannot turn a blind eye to what contractors and subcontractors are doing on site and must accumulate evidence to show that you understand your responsibilities and undertake checks and reviews of works carried out.

Establishing regular project review meetings with representatives of all contractors and specialists would be a good first step. Designers and the Client should be involved. Do not assume that someone else will arrange the meetings. Be proactive.

Understand fully the specification of works and what is expected to be delivered to the Client at the end of the project.

Understand fully the relevant Building Regulations, the CDM Regulations 2015 and any other specialist legislation – for example, environmental management, waste controls, pollution laws, and so forth.

If you are unsure, appoint specialists to the team.

Keep detailed records of all discussions and confirmation that the regulations have been complied with.

Involve construction materials suppliers and question the products being supplied to site and whether they meet agreed specifications. Again, do not take matters on trust – check quality and standards against known criteria.

Participate in design team meetings and offer alternative solutions to design issues raised or alternative solutions which can demonstrate best practice.

Foster collaboration and a non-competitive culture on site.

Keep project plans up to date and realistic time frames.

Report progress, problems, lack of information and potential setbacks to the project team and especially to the Client.

Collate all available information from the beginning of the project and set out expectations for the type and quality of information required at the end of the project.

Remember: CDM Regulations 2015 need a Health and Safety File and higher-risk buildings need a Safety Case Report.

Formalise a process to review contractors and others and have a system whereby concerns can be raised about ongoing competency standards.

Introduce robust contractor and specialist trades competency assessments and ensure that others who appoint contractors have systems in place and can demonstrate that relevant compliance for the project in question is assessed.

What is the Construction Control Plan that is required to be submitted to the Building Safety Regulator in respect of higher-risk buildings?

Certain documents must be prepared and submitted as part of the building control application for higher-risk buildings. These documents must be updated throughout construction as they form part of the information that must kept in relation to the building and handed over to future building owners.

Copies of the documents can be uploaded to the Building Safety Regulator website as part of the application and the original documents kept on behalf of the Client.

The Construction Control Plan must set out details for the work and how you are going to control the work to comply with the Building Regulations. It must describe how your strategies, policies and procedures will make sure

- the building works will meet the Building Regulations
- that you check the plan is effective and how you will record evidence that the building works continue to meet the Building Regulations
- you capture 'as-built' evidence to support the Completion Certificate application
- those carrying out the work are competent to perform their duties
- sanctions and past conduct of those carrying out the work is considered to decide if they are suitable for the role, or how to prevent a repeat of the behaviour
- designers, contractors and others working on the building cooperate and share information
- those carrying out the work have enough information and training.

It must also give the name and a summary of the responsibilities of

- the Principal Contractor
- the Principal Designer
- any other organisation or sole trader who will carry out the work.

The Construction Control Plan must describe how and when you will review the plan.

Is the Construction Control Plan the same as the Construction Phase Plan required under Construction (Design and Management) (CDM) Regulations 2015?

No, they are two different documents required under different legislation.

The Construction Phase Plan required by the CDM Regulations 2015 is concerned with how health and safety will be managed on site. It must be prepared by the Principal Contractor before works commence on site.

The Construction Control Plan is required under the Building Safety Act 2022 and its secondary legislation and must address the topics set out in the question above.

Key elements of a Building Safety Act Construction Control Plan template

Project details

Project name, location, building type, relevant building regulations and designated 'duty holders' (Client, Principal Designer, Principal Contractor).

Higher-risk building identification

Clearly state why the building is considered higher risk, based on factors like height, occupancy or complex design.

Golden thread information management

Detailed procedures on how to gather, maintain and update crucial building information throughout design and construction, including details of materials, design specifications and compliance records.

Risk assessment and mitigation strategies

Identify potential hazards associated with the construction work and outline specific control measures to manage them.

Construction phase management

Stage-specific plans for managing construction activities, including sequencing, access control and safety procedures.

Procedures for managing design changes and variations, ensuring compliance with the Building Regulations.

Monitoring and inspection protocols to verify compliance with design and construction standards.

Communication and collaboration

Clear communication channels between all duty holders, including designers, contractors and building control authorities.

Procedures for reporting incidents, non-compliances and safety concerns.

Competency and training requirements

Ensure all personnel involved in the project have the necessary skills and qualifications to manage building safety risks.

Building Regulations Compliance statement

Confirmation that the design meets Building Regulations and reasons the approach was adopted.

The Building Safety Act 2022: Questions and Answers: A practical approach to design and management of higher-risk buildings

Pat Perry
ISBN 978-1-83662-467-7
https://doi.org/10.1108/978-1-83662-464-620251013

Chapter 11
Fire safety in high-rise residential buildings

What are some of the lessons relating to fire safety in high-rise buildings that have been learned so far from the Grenfell Tower fire?

The Phase 1 Report of the Public Inquiry, led by Sir Martin Moore-Bick, contained a number of recommendations that the government has committed to adopting in full.

In summary, the key recommendations and learnings are as follows.

- The owner and manager of every high-rise residential building be required by law to provide their local fire and rescue service with information about the design of its external walls together with details of the materials of which they are constructed and to inform the fire and rescue service of any material changes made to them.
- All fire and rescue services ensure that their personnel at all levels understand the risk of fire taking hold in the external walls of high-rise buildings and know how to recognise it when it occurs.
- The owner and manager of every high-rise residential building be required by law to provide their local fire and rescue services with up-to-date plans in both paper and electronic form of every floor of the building identifying the location of key fire safety systems.
- The building contains a premises information box, the contents of which must include a copy of the up-to-date floor plans and information about the nature of any lift intended for use by the fire and rescue services.
- All fire and rescue services be equipped to receive and store electronic plans and to make them available to incident commanders and control room managers.
- The owner and manager of every high-rise residential building be required by law to carry out regular inspections of any lifts that are designed to be used by firefighters in an emergency and to report the results of such inspections to their local fire and rescue service at monthly intervals.
- The owner and manager of every high-rise residential building be required by law to carry out regular tests of the mechanism that allows firefighters to take control of the lifts and to inform their local fire and rescue service at monthly intervals that they have done so.
- The owner and manager of every high-rise residential building be required by law to draw up and keep under regular review evacuation plans, copies of which are to be

provided in electronic and paper form to their local fire and rescue service and placed in an information box on the premises.

- All high-rise residential buildings (both those already in existence and those built in the future) be equipped with facilities for use by fire and rescue services enabling them to send an evacuation signal to the whole or a selected part of the building by means of sounders or similar devices.
- The owner and manager of every high-rise residential building be required by law to prepare personal emergency evacuation plans (PEEPs) for all residents whose ability to self-evacuate may be compromised (such as persons with reduced mobility or cognition).
- The owner and manager of every high-rise residential building be required by law to include up-to-date information about persons with reduced mobility and their associated PEEPs in the premises information box.
- In all high-rise buildings, floor numbers be clearly marked on each landing within the stairways and in a prominent place in all lobbies in such a way as to be visible both in normal conditions and in low lighting or smoky conditions.
- The owner and manager of every residential building containing separate dwellings (whether or not it is a high-rise building) be required by law to provide fire safety instructions (including instructions for evacuation) in a form that the occupants of the building can reasonably be expected to understand, taking into account the nature of the building and their knowledge of the occupants.
- The owner and manager of every residential building containing separate dwellings (whether or not they are high-rise buildings) carry out an urgent inspection of all fire doors to ensure that they comply with applicable legislative standards.
- The owner and manager of every residential building containing separate dwellings (whether or not they are high-rise buildings) be required by law to carry out checks at not less than 3-monthly intervals to ensure that all fire doors are fitted with effective self-closing devices in working order.
- All those who have responsibility in whatever capacity for the condition of the entrance doors to individual flats in high-rise residential buildings, whose external walls incorporate unsafe cladding, be required by law to ensure that such doors comply with current standards.

The Phase 1 Report contained a number of additional recommendations addressing fire and rescue procedures, command control systems and communication, firefighter training and other matters.

How does the Grenfell Tower fire affect fire risk assessments for residential buildings, especially those that are high rise?

The Regulatory Reform (Fire Safety) Order 2005 does not apply to domestic premises unless part of the premises is used as a workplace.

The Regulatory Reform (Fire Safety) Order 2005 does, however, apply to the common parts of any residential building – that is, staircases, lobbies, corridors, plant rooms, roof terraces and external areas.

The responsible person – that is, the owner, landlord or managing agent for the residential building – must carry out a fire risk assessment of the common parts.

The Grenfell Tower fire has strengthened the need for fire risk assessments of the common parts and has identified the need to assess external cladding and other building materials as part of the fire risk assessment.

More emphasis needs to be put on the integrity of the structural elements of a building and these should be included in the fire risk assessment.

Fire stopping measures need to be checked to ensure that they have not been breached, especially where refurbishment works have been carried out.

The Grenfell Tower Inquiry has identified that the doors to individual accommodation must also be regularly assessed to ensure that they have the correct standard of fire protection – that is, that they are effective fire doors. Many residents – whether owners, leaseholders or tenants – change their front doors in order to 'personalise' their property. These replacement doors may not have the appropriate fire resistance rating required.

Responsible persons must ensure that signage is clearly visible to residents and users of the building and that means of escape are available to all persons including those with disabilities.

What is a higher-risk building under the Building Safety Act 2022?

Exact wording in the Act states that

> "higher-risk building" means a building in England that—
>
> (a) is at least 18 metres in height or has at least 7 storeys, and
> (b) contains at least 2 residential units.

The Secretary of State may make further Regulations supplementing this section.

The definition in the Act of residential unit is

> "residential unit" means—
>
> (a) a dwelling, or
> (b) any other unit of living accommodation.

Therefore, in addition to residential buildings, care homes and hospitals can also be considered higher-risk buildings if they are higher than 18 m or have at least seven storeys, and so would fall under the new regulatory regime.

The Higher-Risk Buildings (Descriptions and Supplementary Provisions) Regulations 2023 list exclusions to the general definition of higher-risk buildings given above.

For the purpose of Part 4 of the Building Safety Act 2022, buildings are not higher-risk buildings if they comprise entirely of

(i) a care home;
(ii) a hospital;
(iii) a secure residential institution;
(iv) a hotel
(v) military barracks;

or

a building that contains living accommodation for the purposes of the armed forces; or

a building that contains living accommodation for the purposes of any visiting force or an international headquarters or defence organisation designated for the purposes of the International Headquarters and Defence Organisations Act 1964.

The Higher-Risk Buildings (Descriptions and Supplementary Provisions) Regulations 2023 came in to force on 6 April 2023. These Regulations determine which buildings will be subject to the new more stringent safety regime established by the Building Safety Act 2022 ('the 2022 Act'). They do so by supplementing and completing the definitions of 'higher-risk building' at section 120D of the Building Act 1984 and section 65 of the 2022 Act. The 1984 Act makes provision about the design of and building work carried out in respect of higher-risk buildings. The 2022 Act makes provision about safety management in occupied higher-risk buildings.

What are the recommended methods for carrying out a fire risk assessment in high-rise residential buildings?

As of March 2025, definitive government guidance on completing fire risk assessments in high-rise residential buildings has not been issued and the document "Fire Safety in Purpose-built Blocks of Flats", issued by the Local Government Association (LGA) in 2011 and updated in 2020, is still considered applicable guidance. An updated guidance document is expected in 2025.

Where a detailed assessment of external walls of existing multi-storey, multi-occupied residential buildings is deemed necessary, it should now be carried out in accordance with the more comprehensive and holistic guidance included in Publicly Available Specification (PAS) 9980.

Following the Grenfell Tower Inquiry, PAS 9980 has been developed by the British Standards Institution drawing on expert advice from professionals across industry and followed a rigorous development process, including a public consultation.

PAS 9980 provides new guidance on how to assess the risk of fire via an external wall of an existing multi-storey, multi-occupied residential building. PAS 9980 sets out steps that can be taken to identify and assess risk factors as well as mitigation steps that might improve the risk rating of a building via a holistic and fact-based assessment of a building's construction.

Where it is determined that a detailed assessment of an external wall is required, PAS 9980 should now be used for these assessments. It does not contain 'off the peg' solutions to specific wall types and materials but is intended to enable a consistent approach to evaluating the fire risk when considering the external walls of buildings.

There are four types of fire risk assessment for high-rise residential buildings depending on the complexity of the building and knowledge of the building materials. Since the Grenfell Tower fire, it is imperative that owners, landlords, property managers, and so forth, consider the risks of fire spread from external wall cladding and insufficient fire doors to individual dwellings. There is also a need to assess any insulation used within the external framework of buildings as a risk, as it is now accepted that fire can spread via insulation and 'jump' from storey to storey.

Type 1 – common parts only (non-destructive)

A Type 1 fire risk assessment is the basic fire risk assessment required for the purpose of satisfying the Regulatory Reform (Fire Safety) Order 2005.

The inspection of the building is non-destructive. But, as well as considering the arrangements for means of escape, the fire risk assessment includes examination of at least a sample of flat entrance doors. It also considers, so far as reasonably practicable, the separating construction between the flats and the common parts without any opening up of construction.

However, in this type of fire risk assessment, entry to flats beyond the area of the flat entrance door is not involved.

Where there are demountable false ceilings in the common parts, it may be appropriate to lift a sample of readily accessible false ceiling tiles. In addition, it will normally be appropriate to open a sample of service risers, provided access is practicable at the time of inspection.

Unless there is reason to expect serious deficiencies in structural fire protection – such as inadequate compartmentation or poor fire stopping – a Type 1 inspection will normally be sufficient for most blocks of purpose-built flats. Where doubt exists in relation to these matters, the action plan of a Type 1 fire risk assessment may recommend that one of the other types of fire risk assessment be carried out or that further investigation be carried out by specialists.

Type 2 – common parts only (destructive)

The scope and objectives of a Type 2 fire risk assessment are generally similar to those of a Type 1 fire risk assessment, except that there is a degree of destructive inspection carried out on a sampling basis. This will usually necessitate the presence of a contractor for the purpose of opening up construction and making good after the inspection.

In order to check the integrity of separating construction, the areas in which destructive inspection is carried out might sometimes include a sample of flats. However, because of the nature of the work, this can often only be carried out in vacant flats.

A Type 2 fire risk assessment is usually a one-off exercise, which is carried out only if there is good reason to suspect serious structural deficiencies that could lead to spread of fire beyond the flat of fire origin. The age of the block alone is not generally sufficient to warrant a Type 2 inspection. The need for a Type 2 fire risk assessment may sometimes be identified in a Type 1 fire risk assessment, but should not simply be recommended as a matter of course.

Type 3 – common parts and flats (non-destructive)

A Type 3 fire risk assessment includes the work involved in a Type 1 fire risk assessment, but goes beyond the scope of the Regulatory Reform (Fire Safety) Order 2005 (though not the scope of the Housing Act). This risk assessment considers the arrangements for means of escape and fire detection (i.e. smoke alarms) within at least a sample of the flats. Within the flats, the inspection is non-destructive, but the fire resistance of doors to rooms is considered.

Measures to prevent fire are not considered unless (e.g. in the case of maintenance of the electrical and heating installations) the measures are within the control of, for example, the landlord.

A Type 3 fire risk assessment may sometimes be appropriate for rented flats if there is reason to suspect serious risk to residents in the event of a fire in their flats. (This might be, for example, because of the age of the block or reason for suspicion of widespread, unauthorised material alterations.) This type of fire risk assessment will not be possible in the case of long leasehold flats, as there is normally no right of access for freeholders.

Type 4 – common parts and flats (destructive)

A Type 4 fire risk assessment has the same scope of work as a Type 3 fire risk assessment, except that there is a degree of destructive inspection, in both the common parts and the flats carried out on a sampling basis. This will usually necessitate the presence of a contractor for the purpose of opening up construction and making good after the inspection.

However, the nature of the work is such that, often, destructive inspection within flats can only be carried out in those that are vacant. This is the most comprehensive fire risk assessment but will only be appropriate in limited circumstances – such as when a new landlord takes over a block of flats in which the history of works carried out is unknown and there is reason to suspect serious risk to residents from both a fire in their own flats and a fire in neighbours' flats.

(NB: The above four types of risk assessment are reproduced from the 2011 LGA Guide referenced above.)

What needs to be assessed regarding fire doors to individual dwellings?

A fire door ensures that, should a fire break out, it can be contained in a 'compartment'. This keeps the fire and smoke trapped for a defined period, allowing time for people to get out and make the fire easier to tackle.

It will not fulfil this function if damaged or propped open.

Signs that might indicate a fire door include things like a blue 'Fire Door' or 'Keep Closed' sign, door closers, intumescent or smoke seals around the edge of the door or the frame.

In blocks of flats, the external door to a flat invariably should be a fire door; this protects the common areas from spread of flame and smoke. Other locations will depend on the risk assessment and fire plan of the buildings; internal doors could well be fire doors depending on the size of the apartment and distance from the flat entrance door.

All fire doors are fire rated. Some are FD30 (providing 30 min protection), FD60 (60 min protection) or higher. There is usually a certification mark (a label or plug) on top of the door if it is a fire door.

Fire doors should be checked regularly, and the more they are used the more frequently they should be checked.

Anyone can spot a dodgy fire door. But if you have legal responsibility for fire safety, call in a professional.

Create a fire door maintenance checklist and schedule, and check all doors in your building.

Only ever replace damaged components with like-for-like. Check the fire certificate. A trained person should be responsible for this maintenance work.

Five-point check for fire doors

(a) Certification: check that the fire door is certified. There will usually be a label or plug on the top or side of the door. If the door does not include a label or plug, there is no evidence of certification; report it to the owner of the building.

(b) Gaps: the gaps around the top and sides of the door should be no more than 3 mm, plus or minus 1 mm, so no greater than 4 mm and no less than 2 mm. To check if doors meet this standard, use a £1 coin as a guide as this is approximately 3 mm wide. In some instances, the gap at the bottom of the door can be as much as 8 mm. If daylight is seen under the door, the gap is likely to be too big. Where gaps are too big, cold smoke and fire will likely travel through the cracks rendering the fire door non-compliant. Any gaps that exceed this should be reported to the responsible person.

(c) Seals: the intumescent seals around the door are designed to expand when they come into contact with heat. This helps to stop the fire from spreading through the cracks and seals the door shut. If the door does not have seals, or if the seals are faulty, report it to the owner of the building or the responsible person.

(d) Hinges: check that all hinges on the door are firmly fixed and that no screws are missing. If a door or the working parts – such as hinges – are not correctly maintained, it may not be capable of offering the necessary protection in the event of a fire.

(e) Closing: a fire door will only offer the necessary protection if it is closed correctly by a door closer or is locked shut as would a store cupboard or riser door. If the door has a door closer fitted, the door can be tested by opening it to about half-way and then letting go. If the door closes fully by itself without catching on the floor or frame, it is fine. If it does not close correctly, report it. A fire door is no use to anyone if it does not close properly or is wedged open.

What is cladding and why is it important for fire safety?

The aluminium composite external cladding on Grenfell Tower caught fire and fuelled the spread of fire up the external facade of the building.

The term cladding, in its most common usage, refers to the outer skin(s) applied to a high-rise building to increase thermal energy efficiency, and/or to improve aesthetics while not adversely affecting weather resistance. The cladding element is non-load bearing, which means it is not structurally integral to the building itself.

Cladding can either be fitted to an existing building of traditional masonry construction or can be incorporated into the design of a new building.

What are the different types of cladding?

There are many different types of cladding available, ranging from the traditional looking brickwork or rendered systems to more modern metallic rainscreen systems or curtain walls made from glass. Cladding panels can be made from a wide variety of materials such as wood, metal, brick or vinyl, and are often coupled with composite materials that can include aluminium, wood, cement, recycled polystyrene, wheat, rice or straw fibres.

The following are the most common types of cladding facades.

- ACM (aluminium composite material) facade panels are made up of two aluminium skins bonded to a non-aluminium core, often to provide thermal insulation. This is the type of facade panel that was used in the cladding on Grenfell Tower, which caught fire in West London in 2017.
- HPL (high pressure laminate) facade panels are made by pressurising layers of wood or paper fibres into a resin, and then bonding them together using heat. This is the type of facade panel that was used on the University of Bolton student accommodation block that caught fire in 2019.

It is often hard to tell what the material makeup of a cladding system is purely by looking at the facade panels – often, further investigative work will need to be carried out.

How is cladding dangerous?

In fire safety terms, cladding can be dangerous in a few ways. Firstly, there is the makeup of the cladding system itself.

If the cladding system in question contains combustible materials, such as expanded polystyrene (EPS) or polyurethane (PUR), then exposing these materials to a flame will result in fire spread on the outside of a building. This can occur through either a fire breaking out from one of the windows or doorways, through ventilation shafts or from a fire that has occurred on the outside.

Fire-stopping measures are incorporated into nearly every cladding system – however, some are more effective than others. Many are made using mineral-based materials, which are non-combustible and highly effective at slowing the spread of fire around the outside of the building.

Rainscreen cladding systems require an open cavity to allow the insulation to 'breathe' behind the weather-resistant cladding panels. This is potentially dangerous in a fire, as it creates a chimney for the fire to spread vertically through the building. For this kind of cladding to be used, most cavity barriers on the market now incorporate a strip activated by the application of heat, which seals the cavity and when applied correctly dramatically slows the vertical spread of fire. This is called an intumescent cavity barrier.

Is cladding on every building?

No, cladding is not on every building. However, it is becoming increasingly commonly used on both new build high rises and older concrete buildings that have been renovated. That said, just because a building has cladding on it does not necessarily mean that it poses a threat to its occupants. There are many different materials used in cladding systems and many modern materials are perfectly safe from a fire spread perspective.

What do the Building Regulations require regarding cladding?

The Building Regulations were amended in 2018 to implement a ban on combustible building materials on the outside walls of buildings that are taller than 18 m and contain more than one dwelling.

This ban applies to residential housing tower blocks, hospitals, residential care homes and student accommodation blocks.

Fire risk assessments for high-rise tower blocks will also start to include the external walls of a building.

Should a fire test be carried out on all cladding?

That depends on what the cladding is made from. In most cases, records of the materials that have been used in the construction process are held by the building owners or construction contractor – these will give an indication of whether or not the building is at risk.

If this information is not available, a sample of the cladding system may need to be tested in order to determine the materials that it contains, and whether or not there is a potential fire risk.

Owners/managing agents of high-rise buildings are advised to consult PAS 9980:2022 – Fire Risk Appraisal of External Walls and Cladding of Flats, published in 2022. PAS 9980:2022 is a Code of Practice aimed specifically at the safety of existing blocks of flats. Its purpose is to provide a methodology and framework for conducting appraisals of the fire risk posed by combustible external wall construction and cladding.

What changes have the Fire Safety (England) Regulations 2022 introduced regarding external cladding?

As of January 2023, the Responsible Person shall provide information to the local Fire and Rescue Service (FRS) about the design and materials of a high-rise building's external wall system and inform the FRS of any material changes to these walls. They will also be required to provide information in relation to the level of risk the design and materials of the external wall structure gives rise to and any mitigating steps they have taken.

See further on for a template information form.

Fire safety defects in a high-rise block of flats

Lambeth Council was issued an enforcement notice by the London Fire Brigade after firefighters found considerable fire safety defects at a 20-storey tower block.

The failings identified at the council-owned building in Kennington, South London, included, among other things, 'failure to provide and/or maintain adequate and clearly indicated emergency routes and exits that lead to a place of safety' and 'failure to take general fire precautions to ensure the safety of persons on the premises'.

The enforcement notice was issued under the Regulatory Reform (Fire Safety) Order 2005. The London Fire Brigade made the following requirements known, which will need to be complied within a 6-month period.

- Implement the general fire precautions as identified by the fire risk assessment.
- Implement a regular programme to review the fire risk assessment.
- Implement an arrangement for the effective planning, organisation, control, monitoring and review of the preventive and protective measures.
- Ensure that adequate escape routes and exits leading to a place of safety are provided and that these are always maintained clearly and available for use when the premises are occupied.
- Ensure that adequate maintenance systems are in place to ensure that the premises and any facilities, equipment and devices for use by or protection of firefighters are maintained in an efficient state, in effective working order and in good repair.

REFERENCES

BSI PAS 9980:2022 – Fire Risk Appraisal of External Walls and Cladding of Flats.

BSI 9991:2024 – Code of practice for fire safety in the design, management and use of residential buildings.

Building Regulations: Part B 2010. 2019 edition incorporating 2020, 2022 and 2025 amendments and forthcoming 2026 and 2029 changes – for use in England

Building Safety Act 2022.

Fire Safety Act 2021.

Fire Safety (England) Regulations 2022.

Fire Safety (England) Regulations 2022: Factsheets – various. Gov.uk 2022.

Gov.uk: Fire safety in purpose-built blocks of flats, 2011 (due to be updated and republished in 2025).

Regulatory Reform (Fire Safety) Order 2005.

TEMPLATE: EXTERNAL WALL INFORMATION FOR FIRE AND RESCUE SERVICE AS REQUIRED BY FIRE SAFETY (ENGLAND) REGULATIONS 2022

1. Building Identification

1.1. Name, address and postcode of building
1.2. Name and contact details of responsible person

2. Timber Construction

2.1. Are structural timber systems used in the construction of the external walls?	
	Yes
	No

3. Masonry Construction

3.1. Are the external walls constructed from masonry materials?			
	Yes		
	No (go to section 4)		
3.2. Is there any form of cladding or finish present over the outer masonry layer?			
	Yes		
	No (to report additional wall systems go to section 4, otherwise go to section 5)		
3.3. Select external facing materials present over the outer masonry layer			
	Aluminium composite materials		Metal sheet panels
	Other metal composite materials		Render system
	Brick slips		Stone panels
	Glass		Tiling systems
	High pressure laminate (HPL)		Timber
	Other (please specify)		
3.4. Select materials used for insulation between external facing material and masonry layer			
	Mineral wool		Phenolic foam
	Glass wool		Polyisocyanurate (PIR) or polyurethane (PUR) foam
	Expanded polystyrene (EPS) or extruded polystyrene (XPS)		None
	Other (please specify)		

3.5. Are these walls likely to ignite and spread fire easily? Consideration should be given to the combustibility of the external facing material, combustibility of any insulation, and any defects with the design and construction methods (e.g. issues with cavity barriers).	
	Yes
	No (to report additional wall systems go to section 4, otherwise go to section 5)
3.6. Outline the reasons why the walls are likely to ignite and spread fire easily	
3.7. Identify the location of the walls, or sections thereof, which are likely to ignite or spread fire easily. In some instances, the risk of external fire spread will be uniform across a building, in others, the risk will be limited to areas where specific materials have been used (for example, certain floors or elevations).	

If there are additional non-masonry external wall systems to report, then continue to section 4, otherwise go to section 5.

4. Alternative External Wall Systems

Although only presented once below, the questions in section 4 should be answered once for each different external wall system incorporated into the building design – i.e. section 4 may need to be repeated. This is to allow clear differentiation between multiple external wall systems and their associated risk.

4.1. Select the external facing material			
	Aluminium composite materials		Metal sheet panels
	Other metal composite materials		Render system
	Brick slips		Stone panels
	Glass		Tiling systems
	High pressure laminate (HPL)		Timber
	Other (please specify)		

4.2. Select material used for insulation			
	Mineral wool		Phenolic foam
	Glass wool		Polyisocyanurate (PIR) or polyurethane (PUR) foam
	Expanded polystyrene (EPS) or extruded polystyrene (XPS)		None
	Other (please specify)		

4.3. Is this external wall system likely to ignite and spread fire easily? Consideration should be given to the combustibility of the external facing material, combustibility of any insulation, and any defects with the design and construction methods (e.g. issues with cavity barriers).	
	Yes
	No
4.4. If yes, outline the reasons why the walls are likely to ignite and spread fire easily.	
4.5. Outline where on the building this external wall system has been used, and where necessary, how it can be distinguished from the other external wall systems that form part the design.	

5. Wall Attachments and Features

5.1. Does the building include any of the following attachments – Select all that apply			
	Balconies		Photo voltaic panels
	Green walls		Solar shading devices
5.2. Where the attachments selected above are likely to contribute to external fire spread, provide further information below.			

6. Risk and Mitigation

6.1. Following the building's fire risk assessment, was a further fire risk appraisal of the external walls required?	
	Yes, a further fire risk appraisal of the external walls has been completed
	Yes, a further fire risk appraisal of the external walls is required but not yet completed
	No, a further fire risk appraisal of the external walls was not required
6.2. What is the overall level of risk of fire spread due to the design and construction of the external walls?	
	Low risk
	Medium risk
	High
	The overall level of risk of the external walls has not been determined
6.3. What actions have been taken to mitigate the risk relating to the external walls?	
	Change to simultaneous evacuation strategy
	Remediation works to external walls
	Installation of sprinklers
	Removal of gas supply
	No additional measures are necessary

7. Person Completing Report

7.1. Name and contact details of person completing report

(NB: From the government guidance document: Understand your duties as a responsible person under the Fire Safety (England) Regulations 2022.)

The Building Safety Act 2022: Questions and Answers: A practical approach to design and management of higher-risk buildings

Pat Perry
ISBN 978-1-83662-467-7
https://doi.org/10.1108/978-1-83662-464-620251014

Chapter 12
Managing safety risks in occupied high-rise residential buildings

What is a high-rise residential building?

A high-rise residential building has at least

- seven storeys or is at least 18 m high
- two residential units.

The building must be registered with the Building Safety Regulator before people live there.

What must Accountable Persons do to manage fire safety under the Building Safety Act 2022?

The Building Safety Act 2022 says Accountable Persons must take all reasonable steps to

- prevent building safety risks from happening
- reduce the severity of any incidents if they do happen.

The Act defines building safety risks as the risk of spread of fire or structural failure.

What is a reasonable step?

This depends on the building's individual circumstances and risks. The actions taken should be proportionate to the building safety risks and account for the cost and practicality of carrying out the measure.

What are the key steps that need to be taken?

Accountable Persons must

- take all reasonable steps
- check existing safety measures
- set up a safety management system and check and monitor it regularly
- keep records of to the building and refurbishment
- keep information about risk prevention and protective measures
- stop or reduce the likelihood of a risk
- reduce the severity of an incident if one occurs.

How should existing safety measures be checked?

The Accountable Person should try to find and keep a record of all the relevant safety standards which were in place when the building was built and following any later refurbishments.

The Accountable Person should be confident that the building safety measures in place will work as planned. The Accountable Person should be able to prove these were designed, installed, maintained and inspected

- by competent people
- in accordance with relevant legislation, standards and manufacturers' recommendations.

A record of the steps taken to find missing information must be kept as it can be used to demonstrate to the Building Safety Regulator what has been done to find missing information.

If problems with the safety measures are found and they are not sufficient, they should be reviewed and amended. It is important to keep in mind the need to amend information as the building ages and equipment and technical progress develops. Record any changes to maintenance regimes and the reasons for the changes.

If a single measure controls any safety risks, the building is vulnerable if that measure fails. Using more than one measure provides more than one level of protection.

The organisation of the management of building safety risks is as important as the physical measures put in place. Think about planning, organisation, control, monitoring and review. All of these aspects should appear in, and be delivered through, the organisation's safety management system.

If the building undergoes any changes or refurbishment, the Accountable Person should check that they do not affect building safety, both in the refurbished parts of the building and in the building overall.

The Accountable Person should be able to demonstrate that anyone undertaking building work has the necessary skills, knowledge, experience and behaviours (competence) to undertake the designs and construction works.

Changes to other areas of mixed-use buildings can also impact the areas the Accountable Person is accountable for, so information and assurance about these changes will need to be recorded and checked to see if they impact any parts of the building.

During any changes or refurbishment, take the opportunity to see if any additional reasonable steps could be taken as part of the work. For example, an additional measure may now be possible or reasonable due to technical advancements in building safety.

Fire prevention and protective measures often overlap. Accountable Persons should

- be aware of duties under The Regulatory Reform (Fire Safety) Order 2005 (the 'Fire Safety Order')
- cooperate with the responsible person under the Fire Safety Order when considering changes to the building.

Each Accountable Person must keep and update information on prevention and protective measures. This includes

- details of the original design and construction, including drawings
- the current condition of the building
- the types of refurbishment or other changes that have taken place.

Preventative measures aim to stop or reduce the likelihood of a building safety risk happening, especially from fire.

The best way to avoid the spread of fire is to prevent a fire from starting in the first place.

Fire compartments are fire-resisting enclosures providing fire separation between parts of the building. Typical fire compartments include

- individual residential units
- service risers
- staircase enclosures.

In mixed-use buildings, fire compartmentation should also separate residential units from the rest of the building. The purpose of fire compartments is to prevent fire spreading from where it starts to other compartments. Fire compartmentation can also help to protect the means of escape from the building.

Fire compartments, and other fire barriers, must be designed, built and maintained properly to be effective against fire spread.

Effective fire resistance to parts of the building's structure helps to prevent structural collapse if an incident happens.

Fire-resisting door sets, normally referred to as fire doors, help to prevent the spread of fire and sometimes smoke. Fire doors help to protect a building's escape routes and offer fire resistance for a specified number of minutes.

Fire door sets should be specified and installed correctly by a competent person. A fire door set includes the door itself and the frame and door furniture, such as locks and handles.

Check fire doors regularly to ensure they remain in good working condition and will perform as intended in a fire. Check for

- missing or broken self-closing devices
- damaged doors or frames
- doors replaced with non-fire-resisting types.

The Fire Safety Order requires the responsible person for a building to check the fire doors regularly.

Fire stopping provides a seal to prevent the passage of fire where utility services, such as water pipes or electricity cables, penetrate fire compartments. Some fire stops also prevent the passage of smoke.

As well as ensuring initial integrity, pay particular attention to fire stopping during any changes or refurbishment. These may be changes made by you, or by contractors working on the building. They could also be residents making changes to their own flats, such as fitting a new bathroom suite or installing a new broadband connection cable or new front doors.

Any temporary measures needed while work is ongoing should be agreed when planning the work.

Cavity barriers close voids (open spaces) to stop unrestricted movement of smoke and flame. Cavity barriers can be used

- in ceiling voids
- in roof spaces
- in wall cavities
- between floors
- around window reveals.

As with fire stopping, refurbishment or other changes to the building can have an impact on cavity barriers and they are often breached, removed or altered during the works. Changes may affect the integrity of what is already there or need extra cavity barriers. Uninformed contractors can easily destroy cavity barriers when carrying out their work. Ensure contractors working on site understand the importance of such measures and understand what steps need to be taken to maintain their integrity.

A number of measures will help promote structural integrity to prevent a structural building safety risk happening. These include

- initial design and construction
- the competence (skill, knowledge, experience and behaviours) of those involved
- preventing the ingress of water or other substances that might degrade or weaken the building's structure over time
- maintenance and inspection, including acting on any issues found.

Other measures that may prevent or reduce the likelihood of an incident occurring include

- maintenance and inspection of electrical installations in common parts of the building
- lightning protection systems
- measures to prevent or minimise the effect of arson
- gas safety measures
- avoiding the build-up of rubbish.

Protective measures reduce the severity of an incident if one occurs and the Accountable Person is responsible for implementing those necessary in the high-risk building. These can include the following.

Evacuation strategies

Such as

(a) Stay put.
Many high-rise residential buildings are designed to support a 'stay put' approach, sometimes known as 'defend in place'. This means only the people in the compartment where the fire starts (or who are directly affected) are evacuated initially.
People in other areas of the building, who are unaffected by smoke or heat, remain there, safe from the fire. However, they can evacuate if they feel threatened by the fire or are told to leave by the fire service. No-one can be prevented from leaving if they want to.
A 'stay put' policy relies on effective compartmentation, among other measures. Problems with the integrity of compartmentation will affect the viability of a 'stay put' policy.
A stay put policy should only be implemented in consultation with a competent fire safety professional or fire engineer.

(b) Simultaneous evacuation.
Simultaneous evacuation means evacuating the whole building and everyone leaves the building during a fire. This approach requires

- a way of alerting residents in the event of a fire, for example, a fire detection and alarm system
- enough exit routes for all the residents.

The appropriate strategy for each building depends on its original design. It also depends on any subsequent developments that may affect the viability of an evacuation approach.

The Accountable Person must check that the measures designed to support the evacuation strategy being used are in place and working as intended.

People need to know what they should do in the event of an emergency. This is often done using signs, supported by the information required to be given to residents under the Building Safety Act and fire legislation. Residents should also know what to do if they are in a common area, such as a corridor or community room, when a fire breaks out.

Amendments to the Building Regulations Approved Document B (Fire Safety) came in to effect on 1 December 2022 requiring the installation of evacuation alert systems in all new high rise residential buildings taller than 18 metres.

Means of escape

Means of escape are the evacuation routes and exits used by building occupants to reach a place of safety in the event of an incident. It is vital that residents can always use these routes and exits safely. This must be factored into the design and construction of any new building or refurbishment.

Sometimes additional safety measures may be needed such as where there are

- buildings with extended corridors
- mixed-use buildings, for example where residential floors are above office or retail premises
- very tall buildings.

For existing occupied buildings, make sure the means of escape still work effectively. This includes

- general maintenance, inspection and testing
- ensuring corridors are kept clear
- checking fire doors are not propped open or otherwise compromised
- confirming emergency lighting is working.

Travel distances, both within residential units and as part of escape routes, should be in accordance with relevant guidance, or constructed and maintained in line with the building's fire strategy. If these standards cannot be met extra measures may be needed.

Fire detection and alarm

It is not common for high-rise residential buildings, such as blocks of flats, to have a communal fire alarm system on the residential floors. A commercial fire alarm system may be required for common areas used by multiple households like lounges or laundries, or places where people are employed to work (see 'Managing safety risks in high-rise residential buildings: a detailed guide' from Gov.uk/Health and Safety Executive, updated 18 September 2024).

A commercial standard fire detection system can be used to operate active fire safety measures – for example, releasing magnetic locking mechanisms on exit doors, returning lift cars to the ground floor, or operating smoke control and smoke ventilation systems.

Each residential unit, where people live, should have smoke alarms. These are alarms for domestic use and are not usually linked to any other system. Smoke alarms should be checked regularly and changed when their expiry dates are up, as should carbon monoxide detectors where present. Residents' awareness of alarms and detectors should be regularly refreshed through information campaigns, meetings, leaflets, and so forth, so they have the best chance of escaping a fire.

If the building does have a communal fire alarm, it should comply with the correct standard (BS 5839 part 1). Regular maintenance will ensure it works as specified.

Sprinklers

Automatic fire suppression systems, often known as sprinklers, trigger when the temperature in a room reaches 60–70°C. When buildings have sprinklers, they are usually provided within individual residential units rather than common areas.

Sprinklers can tackle a fire soon after it starts, helping occupants escape the burning compartment. They can also help to reduce fire damage.

Prior to 2020, the statutory guidance in England meant that only buildings taller than 30 m and constructed since 2007 were recommended to have a sprinkler system fitted and existing high-rise structures must begin installing sprinkler systems if a fundamental change is made to the structure or use of the building.

In May 2020, the Department for Levelling Up, Housing & Communities published their amendments to the guidance through Approved Document B (fire safety) volume 1. These amendments changed the height requirement for sprinklers by stating that 'blocks of flats with a top storey more than 11 m above ground level should be fitted with a sprinkler system throughout the building.'

Sprinkler systems in accordance with BS 9251 (or BS EN 12845 for residential blocks of flats outside of the scope of BS 9251) should be fitted throughout blocks of flats above 11 m. BS 9251 is the British Standard for the installation of sprinkler systems (among others) that covers

- individual dwellings (houses, flats and maisonettes)
- houses of multiple occupancy (HMOs)
- boarding houses
- blocks of flats
- residential care premises.

Structural integrity

A fire in a building can affect its structural integrity. Structural elements of a building can be protected by cladding them with a fire-resistant material, or by using a protective coating. The condition of the cladding should be checked by regular inspection or surveys.

Other protective measures

These may include

- facilities provided for the fire and rescue services, such as dry and wet firefighting rising mains and firefighting lifts
- smoke control systems
- gas cut offs and utility isolation points
- fire protection to service shafts and risers
- emergency lighting.

The Building Safety Act 2022: Questions and Answers: A practical approach to design and management of higher-risk buildings

Pat Perry
ISBN 978-1-83662-467-7
https://doi.org/10.1108/978-1-83662-464-620251015

Chapter 13
The Safety Case and the Safety Case Report

What is the Safety Case?

The Accountable Person must keep and update certain information about the building and the parts of the building they are responsible for.

The information forms part of the building's Safety Case, which is used to manage the risk of the spread of fire or structural failure.

The Safety Case should be considered as the building's 'information manual', containing information on fire risks and structural risks and the intended procedures or controls to be implemented to reduce those risks for all residents.

The Safety Case should complement the Health and Safety File, prepared under the Construction (Design and Management) Regulations 2015.

Anyone picking up or reviewing documents or files should be able to understand how the building was built and its strengths and weaknesses in respect of life safety risks.

The Safety Case should be regularly updated.

The Principal Accountable Person must use the Safety Case to prepare a Safety Case Report.

What is the Safety Case Report?

The Safety Case Report is a document that summarises the Safety Case for a high-rise residential building. It identifies the building's safety risks and explains how the risks from fire and structural damage are being managed. Building safety risks are the risk of the spread of fire or structural failure.

The Safety Case Report will be submitted at the request of the Building Safety Regulator on applying for a Building Assessment Certificate.

Why is the Safety Case needed?

In her review of fire safety and building regulations, initiated as a result of the Grenfell Tower fire in 2017, Dame Judith Hackett recommended that 'duty holders presented a Safety Case to the JCA (a new Joint Competent Authority) at regular intervals to check that building safety risks are being managed so far as is reasonably practicable'. This has been incorporated into the Building Safety Act 2022.

The Building Safety Act Section 83 introduces new requirements for building owners, in the role of the Accountable Person, to assess and manage building safety risks in high-rise residential buildings. The Act sets out that reasonable measures must be taken to manage the risk of the spread of fire and structural failure, to prevent their occurrence and limit the impact should an incident occur.

Section 84 requires that building safety risks are managed and Section 85 requires that a Safety Case Report is produced and that this is either notified to, or sent to, the Building Safety Regulator.

What are building safety risks?

Building safety risks are defined in Section 62 of the Building Safety Act as

> "building safety risk" means a risk to the safety of people in or about a building arising from any of the following occurring as regards the building—
>
> (a) the spread of fire
> (b) structural failure
> (c) any other prescribed matter.

To date, the government has not indicated that other building risks will be included but if they were so minded, they could add any matter that has the potential to cause a major incident.

> A "major incident" means an incident resulting in—
>
> (a) a significant number of deaths, or
> (b) serious injury to a significant number of people.

What is the difference between a Safety Case and a Safety Case Report?

Safety Cases and Safety Case Reports are complementary elements in ensuring the overall safety of a building. A Safety Case provides the overarching safety argument, outlining the strategies for hazard identification, risk assessment and safety measures, among other critical aspects. It lists all of the information available to the Principal Accountable Person in order for them to fully understand the operation of the building and the design intent for managing building safety risks. A Safety Case is the information that is used to demonstrate that the Accountable Person is taking action to identify, manage and mitigate fire and structural risks, alongside demonstrating what is being done to address these risks.

A Safety Case Report offers detailed evidence and documentation of the safety assurance activities conducted at various stages of the building's life cycle and summarises the key information outlined within the Safety Case, setting out how and what actions and monitoring will be taken to manage the safety risks.

Safety Cases and Safety Case Reports are integral components of ensuring the safety and security of buildings. By effectively integrating Safety Cases and Safety Case Reports, building owners and managers can instil confidence that their structures meet regulatory requirements and protect the well-being of the occupants and the environment.

The Safety Case sets out **what, why and when** you are going to do certain tasks to ensure that building risks are managed.

The Safety Case Report sets out **how you have carried out the checks.** They provide the **evidence** that hazards and risks have been managed.

What should be included in the Safety Case for a higher-risk building?

Include all information relevant to answering the following.

- What could go wrong?
- How could it go wrong?
- Why could it go wrong?

The Accountable Person must compile the Safety Case document for the building. There is no set format as long as the document contains the relevant information. Accountable Persons may want to liaise with the Client for the construction project and discuss with the Principal Designer, appointed under the Construction (Design and Management) (CDM) Regulations 2015, the format of the Health and Safety File and how that might be adapted/enhanced to include information for the Safety Case.

Time and attention should be given to compiling the Safety Case. Once completed comprehensively, it will become a valuable source of information about the building. Do not pad out the document with unnecessary copies of, for example, site construction risk assessments, irrelevant manufacturer's information on services, the construction safety plan, and so on.

The following headings could form the basis of the Safety Case content.

Basic information

Basic information about the building must be included in the Safety Case. This includes, but is not limited to

- the building's height and the number of floors, residential units and staircases
- when the building was built
- relevant design codes or standards
- building control body Completion Certificates (where available)
- plans of the building.

Information about shared facilities, like car parks and a description of the wider area, must also be included, such as

- nearby high-rise residential buildings, schools or care homes
- transport routes like roads or railway tracks
- newer buildings
- buildings built after 2005 that involved more than one contractor will have a Health and Safety File (legislation.gov.uk).

When construction is complete, the owner of the building should keep this in the Health and Safety File required under the Construction (Design and Management) Regulations 2015. If ownership of the building changes, the Health and Safety File must be passed to the new owner.

The building's construction

Information about how the building was built, and the methods and materials used, forms the next section of the Safety Case – for example

- a description of the primary load-bearing system – for example, pre-cast planks on a steel frame
- a description of the stability system – for example, concrete shear walls
- the materials used on the outside of the building
- the roofing material
- the insulation material
- access and escape routes
- a description of the fire compartmentation.

Fire safety strategy

Information on the fire safety strategy for the building must be included. This information will have been previously submitted for new buildings and will be included in the Fire Risk Assessment for older buildings.

Every building is different and information should include

- the Fire Risk Assessment
- fire compartmentation
- fire stopping
- external wall appraisals
- evacuation strategies
- smoke control systems
- firefighter equipment
- access/egress routes
- communication systems.

Refurbishment and other changes

Provide information about any building work or refurbishment works that may affect building safety risks in any part of the building. This can include work on the whole building, common areas or individual parts like residential units – for example

- the Completion Certificate for the refurbishment work
- updated plans of the building
- specification or certification for important materials like replacement cladding.

Resident profile

Information about any residents that can have an impact on building safety risks and emergency plans must be included – for example, residents

- who cannot evacuate without help
- whose first language is not English.

Residents will need to be asked to provide this information. Various ways can be used to collect this information, including by post, email or meetings.

Services and utilities

The building may have utilities like electricity, water, gas, telephone and internet. Information about all these services should be included – for example

- where the supply enters the building
- where and how it can be isolated
- the name and contact details of the supplier
- what the supplier's responsibility for maintenance is
- who carries out the maintenance and repairs.

Building plans should show, for example

- plant rooms and incoming supplies
- pipes and cable routes.

The building may have cables, ducts and pipes that pass through fire compartment barriers. If this is the case, owners or those responsible for the building will need to get assurance that fire stopping has been completed to an appropriate standard. Various sources of information can be checked – for example, from specifications, certifications and surveying examples.

Any services close to the building, like an electrical substation nearby or communications equipment on the roof, must be included in the Safety Case. Even if, as property owner/manager, there is no control or access to these areas, information about how they can affect the building's safety should be included in the document/file.

Assessing risks

Assess the building's safety risks for the parts of the building that you are responsible for. Keep a record of every risk assessment.

Risk assessments must be undertaken

- at regular intervals
- when you think the current assessment is no longer valid
- when asked to do so by the Building Safety Regulator.

Keep track of any actions or recommendations from the risk assessment. Continue to track them until they have been resolved. Make sure that you store and maintain any new information.

Structural safety

Information about the building's structural condition is vital in ensuring building safety.

Keep information about

- any previous structural surveys or inspections
- how you will manage any materials or techniques with known challenges like large panel systems
- any existing programmes of inspection or monitoring relating to structural safety.

Managing risks

Describe the measures that have been put in place to prevent building safety risks from happening and reduce the severity of any incident that does happen.

Keep a record of all the safety measures, which includes

- their design and installation
- how they control building safety risks
- their current condition
- how they are managed and maintained
- any impact on them from building alterations or refurbishments.

Safety management system

Accountable Persons must plan, organise, control, monitor and review the measures taken to manage the safety risks in all parts of the building they are responsible for. The way to do this is referred to as a safety management system.

Accountable Persons do not have to follow any particular safety management system. If there are multiple Accountable Persons for a building, all do not have to use the same safety management system, but there should be coordination with each other when necessary.

Missing information

Accountable Persons must endeavour to find all missing information, especially for existing buildings and after any refurbishment works.

Missing building information can be found in

- archived paper-based storage
- planning authority records
- building control records.

Check that any historical documents remain valid, particularly if the building has been refurbished or altered.

Keep a record of the steps taken to find missing information. You can use it to demonstrate what you have done to find missing information to the Building Safety Regulator.

Finding safety issues

When gathering information about the building, you may find problems, for example

- discovering that a control measure thought to be in place is not
- investigations identifying unresolved issues with control measures.

If you find a safety issue, you should record if any work or action is needed, how long it will take, and whether you need to take any additional measures in the meantime.

If any work is not completed when the Principal Accountable Person needs to submit the building's Safety Case Report, you should provide an action plan to show what work needs to be done, how it will be done and the timescale for completion. The action plan should also include details of any interim measures to control the risk until the work is complete.

What should be included in the Safety Case Report for a higher-risk building?

The key with the Safety Case Report is to evidence that the Accountable Person(s) has(ve) considered the major fire and structural hazards/risks and how the risks are being and will be managed and controlled. It should also identify how and when the report will be reviewed. The Safety Case Report can then be used as a tool to show that all reasonable steps have been, and will continue to be, taken to manage building safety risks.

The following information should be considered for inclusion in the Safety Case Report

- details of the Accountable Persons
- who prepared the report
- building description
- risk assessments summary
- managing risks summary
- safety management system
- planning for emergencies
- ongoing work and building improvement.

A more comprehensive list of information will consist of the following.

Section 1: key information

Provide information about each Accountable Person and, where one is appointed, the Principal Accountable Person, including their

- name
- email address
- telephone number
- postal address in England or Wales.

You should also provide the details of the Responsible Persons for the building, under the Fire Safety Order.

Provide the details of the individuals who prepared the report, such as

- their names
- their job roles and the organisation they work for
- how their skills, knowledge and experience make them suitable for working on the report
- the parts of the report they are responsible for.

Section 2: a description of the building

Your report should include the building name, full postal address, and a photograph or map of where it is.

If your building is part of a wider development, describe it and any shared facilities briefly. Include a description of the nearby area, including other buildings and transport routes.

If there is any missing building information, you should explain what action has been taken to find the missing information.

Safety Case Reports should describe buildings in full, detailing all items the Building Safety Regulator might find relevant. These include the materials used in construction (demonstrating that they are compliant with the latest safety standards), the number of occupants the building has, its fire and structural safety systems, and how safety has been designed and incorporated into the fabric of the building.

The Building Safety Act demands the Accountable Person gather all relevant information about a building via the 'golden thread of information'. In principle, anyone should be able to view these data at any time and see the full specification and history of the building in question.

Some examples of basic building information to include are

- the number of floors, staircases and storeys
- the date the building was built
- the design codes and standards used to build it
- the height of the building

- the type of units the building contains
- a list of communal areas, such as decks, lobbies, stairwells and toilets
- information about the building's surroundings if it is part of a larger development complex
- information regarding nearby transport routes.

Include information relating to the building's construction, such as

- the regulations that were in force when it was built
- how people can escape the building in the event of an emergency
- the types of construction materials used for insulation, cladding, wall attachments and roofing
- the building's stability system
- the building's primary load systems (i.e. steel frame).

If structural systems vary in different parts of the building, you will need to provide details about these separately. You will also need to tell the Building Safety Regulator where one system ends and another begins.

Section 3: information about who will live in the building relevant to safety risks

Safety risks vary depending on the type of residents a building will contain.

Evidence will need to be provided in the Safety Case Report that you are taking steps to mitigate risks. For instance, you may need to list accessibility features that pose a potential hazard, such as electric charging stations for mobility vehicles.

Residents at different stages of their occupancy may need different types of engagement and information. For example, new residents will need to understand their duties when living in a high-rise building.

A deeper understanding of the type of residents living in the building will show that you have a complete view of the building's safety profile.

Section 4: emergency plans and preparedness procedures

All high-rise residential buildings must have preparedness plans detailing how occupants can respond to various emergency events under the new legislation. Therefore, Safety Case Reports must include detailed information on how the building can help residents remain safe in a fire or another emergency.

Section 5: an assessment of the risks residents could face

A risk assessment should contain a list of all the potential hazards building occupants and other nearby stakeholders face. They should detail the likelihood of risks occurring, their impact and their financial ramifications.

When building up a Safety Case Report, you should list and describe the specific building safety risks that residents could face. The process needs to identify

- how things might go wrong in your building
- other things that might make these worse
- potential consequences
- relevant control measures or other mitigation.

Section 6: fire prevention measures

The Building Safety Regulator requires that Safety Case Reports contain fire prevention and protection details. What you need to include will depend on the building's size, location and age.

Most fire prevention and protection should be available via your building's design and specification information. You may also find it in an accompanying Fire Strategy document or existing Fire Risk Assessments.

The first step is to identify the measures already in place. You will then need to provide evidence that all fire mitigation components are installed correctly and working as intended.

Reference to the Fire Risk Assessment will be essential.

The Safety Case Report will need to include evidence that all fire safety prevention and control measures are being checked and regularly maintained.

Section 7: structural safety

The Building Safety Act 2022 considers structural integrity issues a risk for high-rise buildings. Therefore, information about the building's structure and its relation to safety should be included. For example

- your building's foundations and load-bearing systems
- data collected from previous surveys and inspections of structural elements
- the location of secondary structural systems, such as cladding support
- horizontal and vertical stability systems
- identified structural safety issues
- significant changes to the building's structure, perhaps due to renovations.

Section 8: services and utilities information

Multi-occupancy buildings often have different services and utility arrangements from regular, single-unit residences. Furthermore, utility infrastructure is often more complex. Therefore, Safety Case Reports should include

- how electricity, gas, water and internet enter the building
- locations for where operatives can isolate supplies (such as a stop-cock for mains water)
- the names of companies providing installation of utility features and services
- the names of people responsible for utility maintenance
- the location of plant or facility rooms.

Section 9: inspection and maintenance

Information must be provided on who maintains and inspects the building's safety infrastructure, such as fire doors, alarms and sprinklers. This information helps demonstrate that the building remains safe post-construction.

Include information on what gets inspected, how often and by whom.

Include information on how remedial actions are implemented and what post-improvement checks are carried out.

Section 10: refurbishments and future works

Include information on how all refurbishment works and building fabric improvements will be assessed and managed for building safety risks. Many safety procedures and precautions are negated when building repairs and upgrades are carried out as contractors fail to appreciate the measures in place and/or are not supervised sufficiently during re-instatement works – for example, fire stopping is often breached during ducting upgrades and the residual works leave the fire safety integrity of the building breached.

Can I be prosecuted for failing to comply with the Building Safety Act 2022?

Yes. There are numerous offences contained in the Building Safety Act 2022 and penalties include both fines and custodial sentences.

The Principal Accountable Person and the Accountable Person could commit offences regarding the Safety Case and the Safety Case Report as it is they who have the legal duty to produce the documents.

> Section 101: Building Safety Act 2022
>
> Offence: contravention giving rise to risk of death and serious injury
>
> (1) An accountable person for a higher-risk building commits an offence if—
>
> (a) without reasonable excuse, the accountable person contravenes a relevant requirement, and
> (b) the contravention places one or more people in or about the building at critical risk.
>
> (2) In subsection (1)—
>
> "critical risk" means a significant risk of death or serious injury arising from a building safety risk.
>
> "relevant requirement" means any requirement on an accountable person under, or under regulations made under, this Part except one that is prescribed for the purposes of this section.

(3) A person guilty of an offence under this section is liable—

(a) on summary conviction, to imprisonment for a term not exceeding the maximum summary term for either-way offences or a fine (or both);
(b) on conviction on indictment, to imprisonment for a term not exceeding two years or a fine (or both).

and (in either case) is liable on summary conviction to a further fine not exceeding level 1 on the standard scale for each day on which the default continues after the initial conviction.

Section 161: Building Safety Act 2022
Liability of officers of body corporate etc.

(1) Where an offence under Part 2 or 4 committed by a body corporate—

(a) is committed with the consent or connivance of any director, manager, secretary or other similar officer of the body corporate, or any person who was purporting to act in any such capacity, or
(b) is attributable to any neglect on the part of any such person,

that person as well as the body corporate commits the offence and is liable to be proceeded against and punished accordingly.

(2) In subsection (1) "director", in relation to a body corporate whose affairs are managed by its members, means a member of the body corporate.
(3) Subsection (1) also applies in relation to a body that is not a body corporate, with the substitution for the reference to a director of the body of a reference—

(a) in the case of a partnership, to a partner;
(b) in the case of an unincorporated body other than a partnership—

(i) where the body's affairs are managed by its members, to a member of the body.
(ii) in any other case, to a member of the governing body.

(4) Where—

(a) a relevant company is an accountable person for a higher-risk building (within the meaning of Part 4), and
(b) one or more (but not all) directors of the relevant company have been appointed for a building safety purpose and are entitled to remuneration from the company,

this section, so far as relating to Part 4, does not apply in relation to a director who is not entitled to remuneration from the relevant company.

What are the levels of fines and what are the lengths of custodial sentences?

Levels of fines and lengths of custodial sentences will depend on whether the offence is prosecuted in the Magistrates' Court or the Crown Court.

If a Principal Accountable Person fails to do the following without reasonable excuse, they will have committed an offence and will be liable to a fine and/or imprisonment of up to 2 years if found guilty.

- Register a higher-risk building with the Building Safety Regulator before it is occupied (for existing and occupied higher-risk buildings registration must have taken place before 1 October 2023).
- Apply for a Building Assessment Certificate within 28 days of the Building Safety Regulator's direction requiring them to do so.
- Display specified information and documents in a conspicuous position in the building.

If an Accountable Person fails to do the following without reasonable excuse, they will also have committed an offence and will be liable to a fine and/or imprisonment of up to 2 years if found guilty.

- When they cease to be responsible for the building, give prescribed information and a copy of any prescribed document to the Building Safety Regulator and the new Accountable Person.
- Carry out the remedies specified in a compliance notice issued by the Building Safety Regulator on contravention of the requirements on Accountable Person in Part 4 of the Building Safety Act such as assessing and managing building safety risks, developing a Safety Case Report, residents' engagement strategy and complaints procedures, and so on.

In addition, if an Accountable Person fails to do the following without reasonable excuse, they will also have committed an offence and will be liable to a fine if found guilty.

- Give prescribed information on a building safety risk to the Building Safety Regulator by the prescribed time in a specified way.

Legislation

Building Safety Act 2022.

Secondary legislation for higher-risk buildings

- The Building (Higher-Risk Buildings Procedures) (England) Regulations 2023
- The Higher-Risk Buildings (Descriptions and Supplementary Provisions) Regulations 2023
- The Building Safety (Registration of Higher-Risk Buildings and Review of Decisions) (England) Regulations 2023
- The Higher-Risk Buildings (Key Building Information etc.) (England) Regulations 2023

- The Building (Public Bodies and Higher-Risk Building Work) (England) Regulations 2023
- The Higher-Risk Buildings (Management of Safety Risks etc.) (England) Regulations 2023
- The Higher-Risk Buildings (Keeping and Provision of Information etc.) (England) Regulations 2024

Guidance documents from the Department for Levelling Up, Housing and Communities and Building Safety Regulator/HSE

- Safety Case for a high-rise residential building
- Preparing a Safety Case Report
- Making buildings safer – Safety Case Toolkit (https://buildingsafety.campaign.gov.uk/Safety-Case-Toolkit-Reasonable-Steps-and-Practical-Examples-Feb-2024.pdf)
- The Role and Responsibilities of Accountable persons and Principal Accountable Persons – Building Safety Campaign UK Government
- Safety in high-rise residential buildings: accountable persons
- Assessing safety risks in high-rise residential buildings: a detailed guide
- Managing safety risks in high-rise residential buildings: a detailed guide
- Safety management systems for high-rise residential buildings
- Applying to register a high-rise residential building
- Operating a complaints system for building safety in a high-rise residential building
- Operating a mandatory occurrence reporting system
- Register a high-rise residential building
- Complain about a building safety risk in a high-rise building.

Safety Case information checklist (Example)

The following list of information can be used to collate the information usually required to complete the Safety Case for the building in question. It is a non-exhaustive list and as every building is different there may be additional information that will be relevant.

Remember, information is only required that demonstrates how **'building safety risks', as defined in the Building Safety Act 2022, are identified and managed.**

In April 2024, Lee Rowley MP, Housing Minister and Philip White, Director of Building Safety, wrote

> The law is clear: the safety case and the safety case report exist to prevent building safety failures (that is the spread of fire and structural failure) that could result in serious loss of life. Any proposed work to produce these should only be justified against that standard. The safety case and safety case report should not cover issues unrelated to the spread of fire and structural failure.

Information relevant to building safety: *attach or reference all documents and records.*
HRB Registration Number: *Record the registration number allocated by the Building Safety Regulator.*
Policies and procedures of the organisation in respect of building safety and maintenance: *List any policies available such as asbestos management, fire prevention and precautions management, frequency of structural surveys, maintenance procedures etc.*
Training records: *How are employees kept up to date on requirements, new legislation, duties and responsibilities etc.*
Details of the building: *Number of storeys, location, facilities, overall state of repair.*

Design specifications: *Include all design specifications, as-built drawings.*
Fire strategy: *Include means of escape, travel distances, fire mitigation strategies, fire precautions, fire prevention policies, common parts, fire risk assessments, engineered solutions and any fire safety modelling reports, compartmentation standards, fire stopping surveys, fire door surveys etc.*
Fire risk assessments: *Include fire risk assessments for all areas e.g. external areas, fuel storage areas, refuse areas, potential for arson, common parts, external cladding, balcony designs etc.*
Details on firefighting equipment: *Include fire extinguishers, locations, sprinkler systems, suppression systems, maintenance arrangements, available signage etc.*
Building / structural surveys: *Include details on load-bearing systems including secondary systems e.g. cladding systems and balcony attachments.*
Details on refurbishments and completion certificates: *Describe all refurbishment works relating to building safety i.e. structural works and fire safety.*
Resident profiles and evidence of PEEPs: *Who occupies the building and where? List vulnerable residents who will need assistance in an emergency.*

Information about services: *Details on gas, water, electricity, heating sources, battery storage, including supplier and maintenance contact details.*
Maintenance information: *Include information relevant to Annex I of BS 9991 2024: Fire safety in the design, management and use of residential buildings: Code of Practice – relating to all structural and fire safety elements to the buildings.*
Details on lifts: *Include all details and copies of specifications of lifts, including firefighting lifts, goods lifts.*
Evidence of information supplied to residents: *Include details of the residents' engagement strategy, minutes of meetings, complaint procedures etc.*
Wayfinding signage: *Detail how people travel around the building, how floor numbers are displayed and where, how information is shared.*
Floor plans
Safety management plan: *Include information on how building safety risks will be managed e.g. regular surveys, audits.*
Emergency procedures for the building: *What plans are in place to evacuate the building? How will emergency services be notified? Who will be called to assist e.g. structural engineers, emergency responders, alternative accommodation provision, Gold Command procedures etc.?*

Safety Case Report: example content

Summary

The purpose of this Safety Case Report is to provide an overview of the safety measures, risk assessments and management systems in place at (*name of building*). The report aims to assess the building's compliance with relevant safety regulations and identify any potential risks or areas for improvement.

Building information

Name:

Location:

Construction date:

Number of floors:

Number of residential units:

Regulatory compliance

(*Name of building*) complies with all relevant building and safety regulations, including but not limited to

- Building Regulations
- Fire Safety Order
- Health and Safety at Work etc. Act 1974.

Fire safety measures

Fire detection and alarm systems:

(*Name of building*) is equipped with a fire detection and linked alarm system installed on each floor, including common areas and residential units. The system is tested (weekly, quarterly and annually), maintained and inspected by certified professionals to ensure proper functioning.

Emergency exits and escape routes:

The building has clearly marked emergency exits and escape routes, providing residents with safe and efficient evacuation paths in the event of a fire or emergency.

Escape routes are well-maintained, free from obstructions, and adequately illuminated.

Fire suppression systems:

(*Name of building*) has no automatic sprinkler system installed.

Structural safety

Structural integrity:

(*Name of building*) has had a comprehensive structural assessment by qualified engineers to assess its stability and integrity.

The assessment confirmed that the building's structure meets the required safety standards, ensuring the safety of its occupants.

Regular inspections and maintenance

The building management conducts routine inspections and maintenance checks on the structural elements of the tower block, addressing any identified issues promptly and effectively.

A maintenance log is maintained to document all inspections, repairs and upgrades, ensuring transparency and accountability.

Risk assessment and mitigation

Fire Risk Assessment:

A thorough fire risk assessment has been conducted, identifying potential fire hazards and implementing appropriate control measures.

The assessment includes a review of fire safety procedures, training programmes, and the identification of high-risk areas within the building.

Mitigation measures

Fire safety training programmes are regularly conducted for residents, ensuring they are aware of evacuation procedures and fire prevention measures.

Weekly inspections and testing of fire safety equipment, such as automatic air vents, are carried out to ensure they are functional and accessible.

Management systems

Building management team:

(*Name of building*) has a dedicated building management team responsible for the day-to-day operation and safety of the building. The team includes trained professionals who oversee maintenance, safety inspections and resident communication.

Safety reporting and incident management

A robust reporting and incident management system is in place to encourage residents and staff to report safety concerns, incidents or near misses promptly. Resident consultation events have been commenced and a specific dedicated email for fire safety concerns has been created. Residents have access to the repairs reporting management system.

Reported incidents are thoroughly investigated by the Housing Manager and corrective actions are taken to prevent future occurrences. These actions are recorded. In conjunction, an annual Fire Risk Assessment review is undertaken and quarterly fire door visual checks along with the annual invasive door surveys are carried out.

Conclusion

Based on the assessment conducted, (*name of building*) demonstrates compliance with relevant building and safety regulations. The building incorporates various fire safety measures, structural integrity protocols, and a comprehensive risk assessment framework. The management team is committed to ensuring the safety of all residents.

Person completing report

Name:

Position:

Organisation:

Date:

The Building Safety Act 2022: Questions and Answers: A practical approach to design and management of higher-risk buildings

Pat Perry
ISBN 978-1-83662-467-7
https://doi.org/10.1108/978-1-83662-464-620251016

Chapter 14 The three gateways

What are the three gateways applicable to higher-risk buildings?

The Building Safety Act introduced a series of critical gateways to enhance safety standards and oversight for higher-risk buildings in the design, construction and occupation phases.

The Building Safety Regulator, the regulator responsible for overseeing building works to higher-risk buildings, needs assurance and evidence that building safety issues have been properly considered at key stages of a project.

Dame Judith Hackitt's Independent Review of Building Regulations and Fire Safety recommended the gateways process for higher-risk buildings. The gateways process is a series of inspections and safety checks that ensure building safety during construction and design.

Why was the gateways process recommended?

In her final report, Dame Judith Hackitt recommended the following.

> A series of robust gateway points to strengthen regulatory oversight that will require duty holders to show to the JCA [Joint Competent Authority] that their plans are detailed and robust; that their understanding and management of building safety is appropriate; and that they can properly account for the safety of the completed building in order to gain permission to move onto the next phase of work and, in due course, allow their building to be occupied.

The purpose of the gateway system was to

- strengthen regulatory oversight
- improve enforcement tools
- increase accountability for Building Regulations compliance
- improve focus on building safety during construction and design.

What are the gateways?

The gateways are three 'gateway' points where those responsible for a building's safety must prove to the new Joint Competent Authority that they are complying with regulations.

The gateways are intended to introduce an element of governance into managing fire safety.

What are the implications of the gateways?

The gateways regime will increase the time and cost of building and maintaining higher-risk buildings and will require a massive upskilling of all those working on affected projects.

Gateway one focuses on obtaining planning permission by demonstrating the proposed design meets necessary safety requirements – in particular, fire safety.

Gateway two involves building control approval for higher-risk buildings and will include rigorous inspections during construction to verify adherence to approved plans. Gateway two is the hold point whereby construction works cannot commence until the application has been approved by the Building Safety Regulator as complying with all relevant Building Regulations.

Gateway three ensures the building is fit for occupation.

These gateways aim to improve accountability, facilitate early risk identification, ensure ongoing compliance, and establish clear documentation throughout a building's life cycle. Failure to meet gateway requirements can result in penalties and project delays, underscoring the Act's importance in prioritising safety for higher-risk buildings.

Gateway one

What are the detailed requirements of gateway one?

Gateway one is designed to make fire safety an integral part of the design process for tall residential developments. It is the first checkpoint in a series of measures under the broader reforms in the Building Safety Act, which seeks to improve fire and structural safety in buildings across England. Gateway one ensures that fire safety is considered early in the design phase, with the gateway one team at the Building Safety Regulator – the approved Building Control Authority for higher-risk buildings – reviewing planning applications and advising local authorities about the fire safety design of new developments.

Gateway one ensures fire safety is built into the design from the start, making it a priority.

As part of the gateway one process, developers must include a fire statement when submitting planning applications for high-rise buildings. This statement outlines the fire safety measures incorporated into the design.

Planning applications involving new or refurbished higher-risk buildings will be submitted to the local Planning Authority. The Planning Authority are required to consult with the Building Safety Regulator as a 'statutory consultee' on the planning application and the Building Safety Regulator, as the Building Control Authority, will review all the information relevant to fire safety and building safety.

The developer must submit a fire statement setting out fire safety considerations specific to the development with the relevant application for planning permission for development that involves one or more relevant buildings.

Relevant buildings

- contain two or more dwellings or educational accommodation
- meet the height condition of 18 m or more in height, or seven or more storeys.

'Dwellings' includes flats, and 'educational accommodation' means residential accommodation for the use of students boarding at a boarding school or in later stages of education (for definitions see article 9A(9) of the Town and Country Planning (Development Management Procedure) (England) Order 2015 as amended by article 4 of the 2021 Order.

What is the fire statement?

Fire statements will support the consideration of information on fire safety matters as they relate to land use planning matters (e.g. site layout, water supplies for firefighting purposes and access for fire appliances). It is the intention that the information provided within a fire statement is focused and concise, specific and relevant to the development, and proportionate to the scale, type and complexity of the proposal.

A fire statement will evidence that thinking on fire safety matters, as they relate to planning, has been incorporated into the planning application.

What information must a fire statement contain?

Fire statements must be submitted on a form published by the Secretary of State (or a form to similar effect) and contain the particulars specified or referred to in the form, which includes information about (non-exhaustive list)

- the principles, concepts and approach relating to fire safety that have been applied to each building in the development
- the site layout
- emergency vehicle access and water supplies for firefighting purposes
- what, if any, consultation has been undertaken on issues relating to the fire safety of the development and what account has been taken of this
- how any policies relating to fire safety in relevant local development documents have been considered.

Fire safety matters are often relevant to land use planning; they impact on the external layout of a site, including the spaces between buildings. Fire statements will be required to include information on the entire development, as set out on the plan, which identifies the land to which the application relates and that has to be submitted with the application.

Detailed information relevant to complying with Building Regulations does not need to be submitted at gateway one as this information will be considered at the application for building control approval that must be obtained before any construction works can start on site.

Is a fire statement required for a change of use application?

Applications for permission for a material change of use of land or buildings will require a fire statement unless the application is for the following

- a material change in use of a relevant building and the material change of use would result in the building no longer being a relevant building
- a material change in use of land or buildings within the curtilage of a relevant building, unless the material change of use would result in the provision of one or more relevant buildings.

Will following the process of gateway one delay the decisions of the Planning Authority?

No, not necessarily.

The Building Safety Regulator has 21 days from receipt of the application for gateway one assessment in which to consider the application and to provide comments back to the Planning Authority.

There is provision for both the Building Safety Regulator and the Planning Authority to alter the 21 days if necessary. Relevant information should be provided to the applicant.

If applications for gateway one are poorly presented without the necessary information required – that is, the standard form for fire safety statement has not been used – then there may be delays in the information being considered as the Building Safety Regulator may need to return documents for further information or clarification.

Developers and applicants must submit quality information that covers all the necessary details for the Building Safety Regulator to be able to properly consider the proposals.

Gateway two

What is gateway two?

The second gateway point is the building control approval application stage. It will occur prior to construction work beginning on a higher-risk building. It will see a rigorous inspection of Building Regulations come into full effect at each stage of the design and construction process.

Developers cannot proceed with construction until the Building Safety Regulator, as the building control authority, has approved the application.

Gateway two provides a hold point where construction cannot begin until the Building Safety Regulator is satisfied that the design meets the functional requirements of the Building Regulations.

This means that plans need to outline exactly how compliance with the functional requirements of the Building Regulations is going to be met – and that they do not rely on unrealistic management expectations.

The regulator will be stepping in earlier in the design process and implementing a robust evaluation.

The higher the quality of an application, the quicker the Building Safety Regulator can make an assessment.

A multi-disciplinary team will be assigned to each submission, and this will include a Registered Building Inspector, and any other specialists required to assess a submission.

Starting building work without approval from the Building Safety Regulator will be an offence.

What information needs to be submitted with the application?

In your application you will need to

- provide a comprehensive outline of the work that will be carried out
- provide information about the building itself and the people involved in the building, such as the Client, Principal Contractor and Principal Designer
- set out standard building information, such as the site location plan and boundaries, and a Fire and Emergency File
- provide a Construction Control Plan (setting out how building work will be managed to maintain Building Regulations compliance), and a Building Regulations compliance statement (including reasons for the approach adopted)
- provide a Change Control Plan to show how any proposed changes will be considered
- where relevant, provide a partial completion strategy.

What are the detailed requirements for gateway two applications?

At gateway two, the Building Safety Regulator requires you to submit the following.

Competence declaration. You, or someone on your behalf, must sign a declaration that you have assessed your current team's competence to discharge their responsibilities and are satisfied that they meet the minimum threshold. You must ask prospective appointees whether they have received a serious sanction in the last 5 years.

If you discover that they have had a serious sanction, consider this information during your decision to appoint. Seek information about the improvements that have been made, additional training of personnel, latest accident and near miss statistics, and so forth.

You must record in writing the steps you took to assure yourself of the appointees' general and role-specific competence, and their ability to plan, manage and monitor their work.

Drawings and plans. You must supply sufficient drawings and plans to show how you will meet the Building Regulations. To make life easier for the Building Safety Regulator so that they can consider your application efficiently, your plans and drawings should be organised so that they can be accurately and easily referenced in other documents in your application.

Building Regulations compliance statement. This is a method statement demonstrating how you intend the scheme to comply with relevant requirements of the Building Regulations, and why you adopted your approach over alternatives.

Change Control Plan and Change Control Log. This is a system for how you intend to control the changes that inevitably happen during the construction phase of a project. Every unplanned change risks compromising compliance with the Building Regulations and deviating from the outcome approved by the Building Safety Regulator and so must be treated with care.

Construction Control Plan. This is your plan for how the construction works will be managed so that, on completion, your higher-risk building complies with relevant requirements. Evidence should be provided on how cooperation and coordination in the project team will be achieved and competence and good communication will be maintained. It should set out strategies for evidencing compliance and a system for controlling changes. As well as helping to control the project, the document has practical value in communicating strategy to everyone on the project team.

The Construction Control Plan should detail

- the work to be carried out
- all the duty holders and what each is responsible for on the project – probably best set out in some form of matrix
- the strategies to encourage cooperation between project team members and others involved in the project
- the arrangements and policies and procedures for ensuring that the project team knows how to operate the various systems, including
    - the system for managing the construction work
    - the system for controlling changes in accordance with relevant requirements
    - the system for managing the golden thread of information – probably best if it conforms to the ISO 19650 series of standards
- the Fire and Emergency File – this document assures the Building Safety Regulator that implications of the design for managing building safety risks by eventual occupants are properly considered up front.

It must describe the strategies clearly, set out why their underpinning design assumptions are reasonable, and explain how their adoption appropriately mitigates the risks for occupants.

Mandatory occurrence reporting plan. This document explains the system for reporting certain occurrences that threaten the safety of eventual users of the higher-risk building. It includes a statement that you are satisfied that the system is in place and that it is sufficient.

The only occurrences mandated are those that present a risk of death or serious injury to a significant number of people and are therefore usually limited to risks from fire or structural failure.

Partial completion strategy (optional). Where you propose occupation of part of the building before the higher-risk building work is fully completed, you must document your strategy for doing so and include it in your initial application. It must demonstrate that the safety of occupants will not be compromised while the construction works continue.

How long does the Building Safety Regulator have to consider the Building Regulations application?

The Building Safety Regulator has 12 weeks to assess an application to build a new higher-risk building, and 8 weeks to assess an application to carry out building work on an existing higher-risk building.

The Building Safety Regulator may request more information or changes to your application when they receive it, which could entail extra time, which must be agreed with you.

They will give you their decision in writing. Approval lasts for as long as you continue to meet their conditions, which includes doing as described in the application and adhering to

- an inspection schedule for the building work
- the documents that you must maintain for the duration of the project
- a suite of requirement notices, if applicable
- an inspection schedule that includes notification points where work will be assessed before that element of the project can progress.

You may ask for any condition imposed by the Building Safety Regulator to be changed, provided you have good reason.

Part applications

The Building Safety Regulator will accept applications for parts of the project if you cannot viably complete the whole project in one go.

The whole project must be worked out fully before the first application is made. The application must include full descriptions of and a works schedule for all the other parts of the scheme.

The descriptions must include enough detail to assure the Building Safety Regulator that the overall project will, on completion, comply with all relevant requirements – for example, if applying for approval in relation to the higher-risk building's foundations and sub-structure, you must also provide the Building Safety Regulator with sufficient information about its superstructure. Assessing the suitability of the former depends heavily on understanding the specifics of the latter.

How much does the gateway two application cost?

The application fee for building control approval for a higher-risk building (including subsequent stage applications) is £189 (as of April 2025) and this is payable when submitting the application.

The Building Safety Regulator charges £151/h (as of April 2025) to review each application.

There may also be additional charges for any costs that the Building Safety Regulator incurs – that is, for liaising with relevant authorities or third parties.

No fee is payable to submit a notice of intent to start work, commencement or inspection.

What does 'starting' works mean?

Once gateway two approval has been given, the Building Safety Regulator must be notified of the intention to start work at least 5 days before doing so. 'Starting' works means undertaking any element of permanent building works as described in the building control application.

Site set-up, demolition, stripping out works or excavation of trial holes or installing test piles will not be considered as 'starting' higher-risk building works.

The Building Safety Regulator must also be notified within 5 days of the work reaching a point at which it is regarded as 'commenced', which is when foundations, any basement level (if there is one) and the structure of the ground floor is completed.

What will happen during construction?

Enabling works (e.g. site set-up, demolition works, stripping out, excavation of trial holes) may be started without building control approval from the Building Safety Regulator.

Approval must have been received and 5 days' notice given to the Building Safety Regulator before starting any permanent construction work.

You must ensure that your project team

- completes work in accordance with your plans and the conditions of the approval
- keeps documents related to managing and monitoring the project submitted in your application to the Building Safety Regulator up to date
- keeps accurate reports qualifying safety occurrences using your mandatory occurrence reporting system
- keeps records of 'controlled' changes in the Change Control Log – this means changes that involve carrying out construction work other than as set out in the current plans, and/or departing from the strategies, policies and procedures described in current documents must be recorded.

Changes to both design and construction materials are needed on site. Does the Building Safety Regulator need to be informed?

You must tell the Building Safety Regulator about two types of controlled changes.

- Major changes – that is, ones that undermine the basis on which building control approval was granted. You must not allow your project team to implement a major change until you have applied to the Building Safety Regulator and secured their approval for it. The default timescale for the Building Safety Regulator to respond is 6 weeks. However, they may take longer, if necessary, in which case they must agree a new deadline with you in writing.
- Notifiable changes – that is, ones that affect compliance with relevant requirements in the Building Regulations. You must notify the Building Safety Regulator first but otherwise do not have to wait for their approval before making the change.

Key points about notifiable and major changes

Notifiable changes

Generally smaller changes that may affect design details or Building Regulations compliance. Examples: changing the layout of a room within a flat, modifying the location of openings for pipes or cables, altering wall tie systems.

Notification is usually made through a formal process, informing the authority of the change before work commences.

Major changes

Significant alterations that could substantially impact the building's structure, safety or overall design.
Examples: increasing the height of a building, altering the structural design, changing the external wall system, making major changes to the building's use.

Requires detailed submission of change control applications to the relevant authority for review and approval before work can start.

The Building Safety Regulator has the power to reclassify changes and may determine what is thought to be a notifiable change to be a major change. It is good practice to discuss options in the first instance with the Building Safety Regulator otherwise delays to the project may occur due to prolonged decision-making processes.

Gateway three

What is gateway three?

On completion of all building work, contractors are required to submit a gateway three application, which includes a comprehensive building safety case, to the Building Safety Regulator. This submission must provide clear evidence that the construction aligns with regulatory requirements and that a high standard of construction quality has been achieved. Moreover, it must be accompanied by signed declarations from the Principal Designer and Principal Contractor, affirming that the building complies with all regulations.

What information must be submitted to support the application for gateway three?

When submitting the gateway three application, the contractor is responsible for providing the key building information such as

- as-built records
- safety case documents
- golden thread information
- signed declarations from the Principal Designer and Principal Contractor that, as far as is practicable, the buildings and structures constructed comply with all relevant Building Regulations.

The information should be in a digital format and copies given to the building owner.

What happens when the information has been submitted to the Building Safety Regulator?

The Building Safety Regulator will undertake statutory consultations as appropriate with the fire and rescue service and review the submitted information forming the application for gateway three.

The Completion Certificate application involves the submitted information forming the application – signed by the Client, Principal Designer and Principal Contractor – confirming that to the best of their knowledge, the higher-risk building, as built, complies with all applicable requirements of the Building Regulations.

For the Completion Certificate to be approved, there must be sufficient evidence to support that what was built reflects the approved plans, and the requirements of the Building Regulations have been met.

Once the Building Safety Regulator receives this information, the application will be processed as swiftly as possible, so as not to cause undue delay.

Only once gateway three has been passed can the new building be registered with the Building Safety Regulator as required by the Building Safety Act 2022. The building can only be occupied after registration is completed.

Collating the final information is an important duty of the Principal Contractor and Principal Designer. How might this be achieved?

It will be an important part of the design and construction planning process to establish the procedures to be adopted for collating and recording all the key information produced on the project.

The recording systems should be designed and agreed early in the project and the information included in the construction control management plan.

The information collected and recorded should include the following.

As-built drawings: a comprehensive set of updated plans reflecting all changes made during construction, including details of installed mechanical, electrical and plumbing systems.

Construction records: documentation of any variations from approved designs, including change orders, non-conformances, and how they were addressed.

Safety Case Report: a detailed analysis of potential building safety risks, mitigation strategies employed, and evidence of compliance with relevant regulations.

Fire safety information: specific details about fire detection and suppression systems, fire escape routes, and compartmentation strategies.

Structural information: as-built structural details, including calculations and load-bearing elements.

Building materials data: information on all materials used in construction, including fire performance ratings and relevant safety certifications.

Commissioning reports: documentation of testing and verification of building systems to ensure they function as intended.

Maintenance and operation manuals: detailed instructions for ongoing maintenance and operation of building systems.

Contact details: information for key personnel responsible for building management and safety.

Is a Completion Certificate automatically issued by the Building Safety Regulator?

No, not necessarily.

A detailed application informing the Building Safety Regulator that the building works have been completed must be submitted.

For the Completion Certificate to be approved, there must be sufficient evidence to support that what was built reflects the approved plans and the requirements of the Building Regulations have been met.

If the Building Safety Regulator determines that all of the relevant information has not been submitted or that the quality is poor, or that the consultation with other regulatory bodies confirms that information is inadequate, a Completion Certificate will not be issued until the position has been clarified.

The Building Safety Regulator can make site visits to inspect works in order to confirm that Building Regulations have been met.

Is there a charge for the Completion Certificate?

Yes. The Building Safety Regulator applies an application charge of £227 and charges £151/h (as of April 2025) for the time it takes them to verify all the information. Any site visits are charged on an hourly basis.

Additional costs for relevant authorities to assist with verifying the submitted information will be charged at the appropriate rate of the authority involved.

What is a partial completion strategy?

A partial completion strategy is required if it is intended to occupy any part of the building while building work continues in new higher-risk buildings.

In existing higher-risk buildings, a partial completion strategy is required while building works are being carried out if the building

- is still occupied
- any new residential unit created by the work is occupied
- people are moved out of part of the building because of the building work, and then reoccupy it.

The partial completion strategy must set out

- how the building was designed to be occupied in parts
- the parts of the building that will be occupied while work continues
- the measures, strategies and policies the owner should follow to manage safety for occupied parts while building work continues
- any assumptions made, such as the condition of the building or the behaviour of the occupants

The Building Safety Regulator can be directed to any relevant drawings and plans that help describe and support the strategy.

Fire statement form from Gov.uk

Application information	
1. Site address line 1	
Site address line 2	
Site address line 3	
Town	
County	
Site postcode (optional)	
2. Description of proposed development including any change of use (as stated on the application form):	
3. Name of person completing the fire statement (as section 15.), relevant qualifications and experience. Guide: no more than 200 words	
4. State what, if any, consultation has been undertaken on issues relating to the fire safety of the development; and what account has been taken of this. Guide: no more than 200 words	

5. Site layout plan with block numbering as per building schedule referred to in 6. (consistent with other plans drawings and information submitted in connection with the application)
Site layout plan is: Choose an item.

The principles, concepts and approach relating to fire safety that have been applied to the development									
6. Building schedule									
Site information				Building information			Resident safety information		
a) block no. as per site layout plan above	b) • block height (m) • number of storeys excluding those below ground level • number of storeys including those below ground level	c) proposed use (one per line)	d) location of use within block by storey	e) standards relating to fire safety/ approach applied	f) balconies	g) external wall systems	h) approach to evacuation	i) automatic suppression	j) accessible housing provided
		Choose an item.		Choose an item.	Choose an item.	Choose an item.	Choose an item.	Choose an item.	Choose an item.
		Choose an item.		Choose an item.	Choose an item.	Choose an item.	Choose an item.	Choose an item.	Choose an item.
		Choose an item.		Choose an item.	Choose an item.	Choose an item.	Choose an item.	Choose an item.	Choose an item.
		Choose an item.		Choose an item.	Choose an item.	Choose an item.	Choose an item.	Choose an item.	Choose an item.
		Choose an item.		Choose an item.	Choose an item.	Choose an item.	Choose an item.	Choose an item.	Choose an item.
7. Specific technical complexities Explain any specific technical complexities in terms of fire safety (for example green walls) and/or departures from information in building schedule above Guide: no more than 500 words									

8. Issues which might affect the fire safety of the development Explain how any issues which might affect the fire safety of the development have been addressed. Guide: no more than 500 words
9. Local development document policies relating to fire safety Explain how any policies relating to fire safety in relevant local development documents have been taken into account. Guide: no more than 500 words
Emergency road vehicle access and water supplies for firefighting purposes
10. Fire service site plan Explanation of fire service site plan(s) provided in 14. including what guidance documents have informed the proposed arrangements for fire service access and facilities? Guide: no more than 200 words
11. Emergency road vehicle access Specify emergency road vehicle access to the site entrances indicated on the site plan Guide: no more than 200 words Is the emergency vehicle tracking route within the site to the siting points for appliances clear and unobstructed? Choose an item.
12. Siting of fire appliances Guide: no more than 200 words
13. Suitability of water supply for the scale of development proposed Guide: no more than 200 words Nature of water supply: Choose an item. Does the proposed development rely on existing hydrants and if so are they currently usable / operable? Choose an item.

14. Fire service site plan Fire service site plan is: Choose an item.	
Fire statement completed by	
15. Signature	
16. Date	Click or tap to enter a date.

The Building Safety Act 2022: Questions and Answers: A practical approach to design and management of higher-risk buildings

Pat Perry
ISBN 978-1-83662-467-7
https://doi.org/10.1108/978-1-83662-464-620251017

Chapter 15
The golden thread of information

Why was the 'golden thread of information' introduced?

The introduction of a golden thread of information for buildings was a recommendation by Dame Judith Hackitt in her report, 'Building a Safer Future', prepared as a result of the Grenfell Tower fire in 2017, to support duty holders in designing, constructing and managing their buildings as holistic systems.

The golden thread involves keeping a digital record of crucial building information – starting from the design phase and continuing throughout the building's life cycle. The government decided that the golden thread of a building's information must be stored digitally, but this can be on multiple systems.

The Construction (Design and Management) Regulations 2015 require the Principal Designer and the Principal Contractor to prepare a document called the Health and Safety File and to hand it over to the building's owner once the building construction is completed. While this document is often completed, it tends to have little value as a comprehensive building manual as it will frequently contain information about the construction phase – for example, construction phase plan, site Risk Assessment and Method Statement (RAMS), technical data which is no longer relevant and so on. The document was rarely updated when refurbishment works and maintenance works were carried out. This meant that crucial information on the building's systems, materials, design intent, and so forth, were not captured and those that needed to consult such information in order to develop safe systems of work for further projects were unable to do so.

What is the purpose of the golden thread?

The purpose of the golden thread is to have the right information in order to understand the building and the steps needed to keep both the building and the people living in it safe.

To ensure that duty holders identify, understand, manage and mitigate building safety risks throughout the building's life cycle, those responsible for building safety must now retain a 'golden thread of information'.

The right people, the right information, the right time.

It will hold the information that those responsible for the building require to

- show that the completed building and any later building work meet the requirements of applicable Building Regulations

- identify, understand, manage and mitigate building safety risks
- prevent or reduce the severity of the consequences of fire spread or structural collapse throughout a building's life cycle.

Having a golden thread means that those people responsible will have easily accessible, reliable, up-to-date and accurate information. Without this information, it is very difficult to manage buildings safely, because the golden thread content is the information a duty holder needs to enable them to both fulfil and demonstrate that they are carrying out their legal duties.

The golden thread comprises two parts: building work and maintenance. This ensures that any modifications or enhancements to the building are accurately documented and enables building owners to proactively identify potential risks or hazards and take corrective action.

It is important that those who are responsible for the building know where up-to-date information is stored and can give access to the people who need it. This includes anyone responsible for maintaining or working on the building and other relevant groups, such as residents and emergency responders.

Is the golden thread of information a legal requirement?

Yes, in respect of higher-risk buildings. The Building Safety Act 2022 places legal duties on duty holders (Part 3) and Accountable Persons (Part 4) to keep and maintain information about a higher-risk building – this information is known as the golden thread of information.

The legal requirements are not consolidated into one piece of legislation. There are seven sets of regulations that underpin the framework and duties.

The Act and Regulations set out

- what information needs to be kept as golden thread information
- how it should be kept
- who is responsible for keeping and updating the information
- how that information from the golden thread will be shared throughout the life cycle of the building, with whom and when this information needs to be shared.

There are three key regulations for the implementation of the golden thread. These detail the requirements for the information that must be kept as golden thread information and how the golden thread should be kept.

(a) The Building (Higher-Risk Buildings Procedures) (England) Regulations 2023 (Statutory Instrument (SI) 2023/909).
(b) The Higher-Risk Buildings (Management of Safety Risks etc.) (England) Regulations 2023 (SI 2023/907).
(c) The Higher-Risk Buildings (Keeping and Provision of Information etc.) (England) Regulations 2024 (SI 2024/41).

A further four sets of regulations for the higher-risk building regime interlink with the above regulations for the golden thread of information.

(a) The Building Regulations etc. (Amendment) (England) Regulations 2023 (SI 2023/911).
(b) The Building Safety (Registration of Higher-Risk Buildings and Review of Decisions) (England) Regulations 2023 (SI 2023/315).
(c) The Higher-Risk Buildings (Key Building Information etc.) (England) Regulations 2023 (SI 2023/396).
(d) The Higher-Risk Buildings (Descriptions and Supplementary Provisions) Regulations 2023 (SI 2023/275).

When does the golden thread start?

The golden thread needs to be created before building work starts and the information must be kept updated throughout the design and construction process (e.g. where – through the change control process – the plans for the building work are changed).

When the building work is completed, the golden thread must be handed over to the Accountable Person(s), who is responsible for the occupied building.

So, remember your building's information must be

- kept digitally
- kept securely
- a building's single source of truth
- available to people who need the information to do a job
- available when the person needs the information
- presented in a way that can be easily used.

Essentially, the golden thread is all about the right people having the right information when they need it.

- Right people – those who require the information to carry out a function.
- Right information – presented in a way that the receiver can use.
- Right time – when the information will add value.

Each golden thread will be individual, bespoke and specific to the building and group of residents and occupants.

Who is responsible for the golden thread?

During the design and construction phase

The duty to keep and manage the golden thread rests with the duty holder. This could be

- the Client
- the Principal Contractor
- the Principal Designer.

During occupation

The Accountable Person is responsible for coordinating the golden thread, keeping it updated and ensuring it is accurate and accessible. And when they cannot find this information, they need to justify why.

The Accountable Person may be an individual, partnership or corporate body and there may be more than one Accountable Person for a building. If there are multiple Accountable Persons, then the Principal Accountable Person will take lead responsibility for the golden thread.

Where the building already exists, the Accountable Person will need to make reasonable enquiries to find the information that allows them to assess and manage the safety risks of fire spread and structural stability.

During refurbishment

When a building is being refurbished, this may involve both duty holders as well as the Principal Accountable Person and Accountable Persons as many buildings will remain occupied during refurbishment.

Accountable Persons must ensure that all information from the refurbishment works that is relevant to building safety is collated digitally and kept within the Safety Case Report. Information about changes to any fire safety or structural safety of the building may need to be shared with the residents.

How will the golden thread be used?

The golden thread will be used to *provide the right information to the right people at the right time.*

In particular, as follows.

- During the design and construction phases to collate, record and provide relevant information about the building.
- During the occupation phase by the Principal Accountable Person and other Accountable Persons as part of their role in using and operating the building; undertaking maintenance, refurbishment or further construction work; and compiling the Safety Case Report.
- To provide residents with accurate information about their homes and to reassure them that their building is being managed safely.
- To provide the emergency services with relevant information about the building so they can respond appropriately when required. Contact details, copies of the building's floor plans, and details of key firefighting equipment should be available in the secure information box.

What is the format required for the golden thread?

There are three sets of regulations that set out the requirements for establishing and updating the golden thread.

- The Higher-Risk Buildings (Management of Safety Risks etc.) (England) Regulations 2023 – Regulation 7.
- The Building (Higher-Risk Buildings Procedures) (England) Regulations 2023 – Regulation 31.
- The Higher-Risk Buildings (Keeping and Provision of Information etc.) (England) Regulations 2024 – Regulation 4 and Schedule 1.

Although different wording is used in the sets of regulations, the key requirements for the golden thread are that it must be

- kept in an electronic format
- capable of being transferred electronically without data being lost or corrupted
- accurate and up to date
- intelligible to the intended readers of the data, with any key needed to understand it provided
- accessible and made available to other duty holders on request
- secure from unauthorised access
- only changed in accordance with procedures that record the person who made the change and the date
- consistent in language, terminology and definitions.

The Regulations do not prescribe any particular format for the golden thread; however, ensuring that relevant information can be easily identified and accessed by the Principal Accountable Person is key to making sure the building is maintained and operated safely.

The information can be kept in a range of locations but it is important that all the relevant information is easily available and therefore the golden thread information will need to be cross-referenced with other documents – for example, the Health and Safety File produced during construction works and required by the Construction (Design and Management) Regulations 2015, the Fire Risk Assessment, and so on.

The government advocates the use of Building Information Management/Modelling Systems.

What should be included in the golden thread?

Schedule 1 of The Higher-Risk Buildings (Keeping and Provision of Information etc.) (England) Regulations 2024 lists the information that must be included in the golden thread.

Information should be clearly laid out to facilitate ease of access to key information. A logical format might include section/chapter headings.

Examples might include the following.

Building registration information

- Name and address
- Number of residential units

- Height of the building
- Number of storeys above ground level
- Date of completion

Building Control Approval information

- Drawings and plans
- Competence declaration
- Construction control plan
- Change control plan and log
- Building Regulations compliance statement
- Fire compliance statement
- Fire and Emergency File
- Client authorisation
- Site location plan including any new drains and sewers
- Drainage precautions
- Arrangements for staged work
- Partial completion strategy where appropriate

Certificates and reports

- Planning statement
- Completion Certificate
- Building Assessment Certificate
- Safety Case Report
- Inspection and test reports, including recommendations and actions taken
- Maintenance records and actions taken
- Mandatory Occurrence Reports
- Complaints received and actions taken (within the last seven years)
- Any enforcement action

Inspection regimes

- Statutory inspections required
- Standards to be met

Maintenance information

- Schedule of planned maintenance for building elements and equipment
- Dealing with faults
- Troubleshooting
- Repair and replacement details
- Competence requirements for maintenance personnel

Building safety management documentation

- Evacuation strategy
- Mandatory Occurrence Reporting systems and processes

As-built drawings

- Up-to-date site location plan, including all services
- Photographic records of building elements and installations during the construction process
- Up-to-date floor plans

Safety-critical information about the building that cannot be compromised

- Design principles
- Structural elements
- Utilities
- Compartmentation
- Fire safety systems (i.e. alarm systems, evacuation strategy and escape routes, fire doors, firefighting equipment, smoke vents)
- Electrical systems
- Ventilation systems
- Window security systems

Product-specific details

- Certification
- Supplier
- Material content

Refurbishment-specific details

- Hidden services – for example, smoke vents
- Passive fire protection – for example, location of fire collars, fire stopping

Demolition-specific details

- Load-bearing elements
- Post-tensioned elements
- Foundation details

Who holds or keeps the golden thread of information?

When the building work is completed, the golden thread must be handed over to the Accountable Person(s), who is responsible for the occupied building. Where there may be a Principal Accountable Person and Accountable Persons responsible for the same building(s) there must be a clear information 'road map' indicating who has what information, where it can be accessed and how it can be accessed.

No-one should keep key building information away from those who will need it.

Remember: the golden thread of information provides *the right information to the right people at the right time.*

What are the golden thread principles?

The Higher-Risk Buildings (Management of Safety Risks etc.) (England) Regulations 2023 Regulation 7 states

> 7.(1) The standards prescribed for the keeping of information under section 88(1) of the 2022 Act are that the information—
>
> (a) is kept in an electronic format which ensures it is capable of being transferred electronically to other persons without the data in it being lost or corrupted;
> (b) is accurate;
> (c) is intelligible to the intended readers of the data, and any key needed to understand the data is provided with the data;
> (d) is kept in such a manner as to be accessible as soon as reasonably practicable in response to a request from any person specified in section 89(1) of the 2022 Act or prescribed by regulations made under that subsection.
> (e) is secure from unauthorised access; and
> (f) is only changed in accordance with procedures which record the person who made the change and the date of that change.
>
> (2) The standards prescribed for the keeping of copies of documents under section 88(2) of the 2022 Act are that the documents—
>
> (a) are kept in an electronic format capable of being transferred to other persons without the data in them being lost or corrupted;
> (b) are kept in such a manner as to be accessible promptly in response to a request from any person specified in section 89(1) of the 2022 Act or prescribed by regulations made under that subsection.
> (c) are secure from unauthorised access; and
> (d) are only changed in accordance with procedures which record the person who made the change and the date of that change.

The Building Regulations Advisory Committee have produced a set of ten **golden thread principles** that, if followed, should ensure that the golden thread meets the expectations of the legislation and ensures that, overall, the safety of buildings is enhanced because relevant information will be available to those who need it.

1. **Accurate and Trusted:** the duty holder/Accountable Person/Building Safety Managers and other relevant persons (e.g. contractors) must be able to use the golden thread to maintain and manage building safety and ensure compliance with building regulations. The Regulator should also be able to use this information as part of their work to assess the compliance with building regulations, the safety of the building and the operator's safety case report, including supportive evidence, and to hold people to account. The golden thread will be a source of evidence to show how building safety risks are understood and how they are being managed on an ongoing basis. The golden thread must be accurate

and trusted so that relevant people use it. The information produced will therefore have to be accurate, structured, and verified, requiring a clear change control process that sets out how and when information is updated and who should update and check the information.

2. **Residents feeling secure in their homes**: residents will be provided information from the golden thread – so that they have accurate and trusted information about their home. This will also support residents in holding Accountable Persons and Building Safety Managers to account for building safety. A properly maintained golden thread should support Accountable Persons in providing residents the assurance that their building is being managed safely.
3. **Culture change:** the golden thread will support culture change within the industry as it will require increased competence and capability, different working practices, updated processes and a focus on information management and control. The golden thread should be considered an enabler for better and more collaborative working.
4. **Single source of truth:** the golden thread will bring all information together in a single place meaning there is always a 'single source of truth'. It will record changes (i.e. updates, additions or deletions to information, data, documents and plans), including the reason for change, evaluation of change, date of change, and the decision-making process. This will reduce the duplication of information (email updates and multiple documents) and help drive improved accountability, responsibility and a new working culture. Persons responsible for a building are encouraged to use common data environments to ensure there is controlled access to a single source of truth.
5. **Secure:** the golden thread must be secure, with sufficient protocols in place to protect personal information and control access to maintain the security of the building or residents. It should also comply with current GDPR legislation where required.
6. **Accountable**: the golden thread will record changes (i.e. updates, additions or deletions to information, data, documents and plans), when these changes were made, and by who. This will help drive improved accountability. The new regime is setting out clear duties for duty holders and Accountable Person for maintaining the golden thread information to meet the required standards. Therefore, there is accountability at every level – from the Client/Accountable Person to those designing, building or maintaining a building.
7. **Understandable/consistent:** the golden thread needs to support the user in their task of managing building safety and compliance with building regulations. The information in the golden thread must be clear, understandable and focused on the needs of the user. It should be presented in a way that can be understood, and used by, users. To support this, duty holders/Accountable person should where possible make sure the golden thread uses standard methods, processes and consistent terminology so that those working with multiple buildings can more easily understand and use the information consistently and effectively.
8. **Simple to access (accessible):** the golden thread needs to support the user in their task of managing building safety and therefore the information in the golden thread must be accessible so that people can easily find the right information at the right time. This means that the information needs to be stored in a structured way (like a library) so people can easily find, update and extract the right information. To support this the government will set out guidance on how people can apply digital standards to ensure their golden thread meets these principles.

9. **Longevity/durability and shareability of information**: the golden thread information needs to be formatted in a way that can be easily handed over and maintained over the entire lifetime of a building. In practical terms, this is likely to mean that it needs to align with the rules around open data and the principles of interoperability – so that information can be handed over in the future and still be accessed. Information should be able to be shared and accessed by contractors who use different software and if the building is sold the golden thread information must be accessible to the new owner. This does not mean everything about a building and its history needs to be kept, the golden thread must be reviewed to ensure that the information within it is still relevant and useful.
10. **Relevant/proportionate:** preserving the golden thread does not mean everything about a building and its history needs to be kept and updated from inception to disposal. The objective of the golden thread is building safety and therefore if information is no longer relevant to building safety it does not need to be kept. The golden thread, the changes to it and processes related to it must be reviewed periodically to ensure that the information comprising it remains relevant and useful.

(Policy paper by Building Regulations Advisory Committee: Golden thread report (published 21 July 2021.)

One of the golden thread principles is the 'single source of truth'. What does this mean?

The golden thread will bring all information together in a single place, meaning there is always a 'single source of truth'.

It will record changes (i.e. updates, additions or deletions to information, data, documents and plans), including the reason for change, evaluation of change, date of change and the decision-making process.

This will reduce the duplication of information (email updates and multiple documents) and is expected to help drive improved accountability, responsibility and a new working culture.

Persons responsible for a building are encouraged to use common data environments to ensure there is controlled access to a single source of truth.

The common data environments should be discussed at the commencement of the project and the Client should set out the expected format of information that will need to be handed over at the end of the project.

What is needed during design and construction of a higher-risk building?

General information about the building should be included in the golden thread and the Client's responsibilities should be clearly set out.

All clients must keep information that describes the building and shows how it complies with Building Regulations. A record of the following should be kept.

- How those responsible for the building work comply with their duties.
- How you will manage the work, so you build what you set out to build.

Information about the building must be shared with every designer and contractor on the project and procedures can be integrated with various duties in the Construction (Design and Management) Regulations 2015.

What role does the Principal Designer have in relation to creating the golden thread?

The Principal Designer must capture and keep all the design information up to date.

The Principal Designer must work with the Principal Contractor to

- make sure the design work during the construction phase is recorded and evidenced in easily accessible documentation
- help the Client to provide information to anyone who needs it
- share information relevant to the building work
- have processes in place to help capture all relevant information
- facilitate communication between designers
- capture and share information relevant to the building work.

What role does the Principal Contractor have in relation to creating the golden thread?

The Principal Contractor must use the record-keeping system provided by the Client to ensure that information about the building is up to date.

The Principal Contractor must

- help the Client to provide information to anyone who needs it
- work with the Principal Designer to share information relevant to the building work
- have processes in place to help communication between all contractors
- keep evidence that the building work meets the Building Regulations
- work with the design team to manage controlled changes
- ensure that design changes are recorded.

What information is needed after a refurbishment project?

The golden thread often fails when refurbishment projects are undertaken because no-one takes responsibility for updating existing information with details of the works carried out.

The Grenfell Tower Public Inquiry found that vital information about the building was not made available to the London Fire Brigade and that no information was available about the recent refurbishment project and the construction materials used, especially in relation to the external cladding. The lack of information hampered the effectiveness of the firefighting and evacuation of the tower block.

The Construction (Design and Management) (CDM) Regulations 2015 require that a Health and Safety File is completed for every building project and handed over to the building owner at the end of the construction phase. The file should be kept up to date with new information from all refurbishment and maintenance works. But, sadly, often it is not. No-one takes responsibility or it is just an overlooked task to add relevant information about the maintenance and building works.

Any information relating to building and fire safety should be added to the Health and Safety File or to any other record-keeping systems containing information about the building and its use.

Information about any works carried out that enhances, alters or improves fire safety must be added; information about structural safety; information about construction materials and specifications; information about changes to emergency procedures; changes to maintenance schedules and building services, security services, and so forth, should all be included.

How do the three 'gateways' referred to in the Building Safety Act 2022 fit into the golden thread?

The Building Safety Act gateways are designed to provide a comprehensive safety framework throughout the life cycle of a higher-risk building. Information supporting each gateway and essential information about the design and intent, construction and future use of a higher-risk building are critical.

Information submitted under each gateway should be included in the golden thread.

There are three gateways.

- Gateway one: the planning gateway: before planning permission is granted.
- Gateway two: the construction phase gateway: before work can start.
- Gateway three: the occupation gateway: before the building can be occupied.

Gateway one requires clients and designers to consider the fire safety needs as they relate to the project. This includes the submission of a full fire statement.

Client (individuals or organisations commissioning the project): submits Fire Statement and planning application.

Principal Designer (individuals or organisations overseeing the design process): prepares safe design.

The Health and Safety Executive has been designated the statutory consultee for higher-risk building planning applications and they will review the Fire Statement and provide advice to the Local Authority on fire safety matters.

The Local Authority is responsible for granting planning permission.

Obtaining approval at gateway one is a prerequisite for proceeding with the construction phase. By addressing safety concerns early on, gateway one aims to prevent costly remediation measures and potential safety hazards further into the construction phase.

Gateway two is the second critical checkpoint in the Building Safety Act's framework, occurring during the construction phase of higher-risk buildings. This gateway focuses on ensuring that the practical implementation of the approved design adheres to safety standards and requirements. There can be no deviations from the original design approved under gateway one, thus safeguarding the integrity of the building's safety features.

Effectively, gateway two provides the stop/go decision before construction.

The Building Safety Regulator will conduct rigorous inspections and assessments to verify that the construction process aligns with the plans and specifications approved during gateway one. They examine the materials, workmanship and installation of critical safety systems, such as fire safety measures, structural elements and safety-critical systems.

The Client (individuals or organisations commissioning the project) must submit evidence of

- compliance with Building Regulations
- provision of full plans
- appointment of competent people
- application for Building Control Approval.

Any alterations to current plans must be documented in a change control log. Under the Building Safety Act, these modifications are categorised as either

- major changes, which necessitate an application to the Regulator and approval before work can start
- notifiable changes, where work can commence once the Regulator has been informed.

Gateway three, the final gateway under the Building Safety Act, is applicable during the occupation phase of a higher-risk building. This final and critical stage ensures that the completed building meets all safety requirements and is fit for occupation.

The Client submits an application for a Completion Certificate, providing evidence that the building has been constructed according to the approved plans. Evidence that must be submitted includes

- demonstrating the building is safe to occupy
- applying for the Completion Certificate
- showing ongoing management plans.

The Principal Contractor must hand over the golden thread of building and operational information that has been collected during the course of the construction.

The golden thread is an accurate and up-to-date record of all information for a building from design through to occupation. The three gateways will support the creation of the golden thread of information.

Clear duties and responsibilities must be allocated at the start of a project, preferably in writing so that there is no confusion as to who is responsible for which task regarding the gathering of information.

In summary the following information should be gathered by

- planning application – Principal Designer
- construction phase – Principal Contractor
- occupation – Accountable Person.

For each of the reporting gateways, a duty holder is required to provide the golden thread of information to the Building Safety Regulator. Through design to construction, the duty holder is the Principal Designer and then the Principal Contractor, and once the building is occupied with residents the responsibility falls on the Accountable Person.

NB: Please refer to Chapter 14: The three gateways.

Legislation

The following secondary legislation, made under the Building Safety Act 2022, relates to higher-risk buildings.

- The Higher-Risk Buildings (Descriptions and Supplementary Provisions) Regulations 2023
- The Building Safety (Registration of Higher-Risk Buildings and Review of Decisions) (England) Regulations 2023
- The Higher-Risk Buildings (Key Building Information etc.) (England) Regulations 2023
- The Building (Public Bodies and Higher-Risk Building Work) (England) Regulations 2023
- The Building (Higher-Risk Buildings Procedures) (England) Regulations 2023
- The Higher-Risk Buildings (Management of Safety Risks etc) (England) Regulations 2023
- The Higher-Risk Buildings (Keeping and Provision of Information etc.) (England) Regulations 2024

Other regulations have been introduced relating to building control, construction products, leaseholder protections and other subsidiary subjects included in the Building Safety Act 2022.

Guidance documents

Various guidance documents have been produced by government departments, the HSE and the Building Safety Regulator. Some have been withdrawn as the Building Safety Act passed through the parliamentary process, but often the information contained is still useful. More guidance documents are under development.

Building a Safer Future (May 2018): Final Report from an Independent Review of Building Regulations and Fire Safety: Dame Judith Hackitt.

Building Safety Regulator/HSE: The three gateways to building safety.

Construction Leadership Council: Delivering the golden thread: Guidance for duty holders and accountable persons (August 2024).

Gov UK: Guidance: Keeping information about a higher-risk building: the golden thread.

Gov UK: Golden thread: Factsheet (This guidance was withdrawn on 25 July 2022).

Gov UK: Making buildings safer: Understanding the golden thread.

HSE: Golden thread webinar (April 2024).

Making Buildings Safer: various webinars and vlogs via HSE website.

The Building Safety Act 2022: Questions and Answers: A practical approach to design and management of higher-risk buildings

emerald PUBLISHING ice Publishing

Pat Perry
ISBN 978-1-83662-467-7
https://doi.org/10.1108/978-1-83662-464-620251018

Chapter 16
Mandatory occurrence reporting

What is mandatory occurrence reporting?

Mandatory occurrence reporting is a requirement to report safety issues in buildings.

Mandatory occurrence reporting is a system introduced under the Building Safety Act 2022 and its subsidiary legislation that principal duty holders must establish and maintain systems to enable prompt reports of information regarding safety occurrences by the reporting person/s to the principal duty holders.

Mandatory occurrence reporting aims to improve the communication of building safety risks with the Building Safety Regulator. Mandatory occurrence reporting has been introduced to increase reports of any structural flaws and fire risks that could lead to catastrophic disasters, and accordingly, mitigate safety risks throughout the life cycle of a building.

Duty holders must operate a mandatory occurrence reporting system during

- the construction phase of a new higher-risk building
- work on an existing higher-risk building or a building that will become a higher-risk building
- occupation of a higher-risk building.

What is a safety occurrence?

A safety occurrence is any incident or situation relating to the structural integrity or fire safety of a higher-risk building that meets the risk condition. It also includes any design aspects relating to the structural integrity or fire safety of a higher-risk building that would, if built, meet the risk condition.

What is a risk condition?

The 'risk condition' is where a part of a building in use would be likely to present a risk of a significant number of deaths, or serious injury to a significant number of people due to an incident or situation not being remedied.

Is a building safety risk the same as a safety occurrence?

The Building Safety Act 2022 Section 62(1) defines a 'building safety risk' as a risk to the safety of people in or about a building arising from any of the following occurring as regards the building

- the spread of fire
- structural failure
- any other prescribed matter.

The Higher-Risk Buildings (Management of Safety Risks etc.) (England) Regulations 2023 defines a 'safety occurrence' as an incident or situation relating to the structural integrity of, or spread of fire in, a higher-risk building that meets the risk condition.

So, in essence they are the same – that is, something that has the potential to cause safety risks to persons occupying higher-risk buildings.

What are some examples of what must be reported as a safety occurrence?

Examples of safety occurrences include, but are not limited to

- the spread of fire (or something that could lead to the spread of fire)
- total or partial collapse of the building
- defective building work
- unexpected failure or the degradation of construction materials
- the discovery of structural defects
- failure of a critical fire safety measure, such as an automatic opening vent, smoke extraction, fire doors or smoke detectors.

Who is responsible for making a mandatory occurrence report to the Building Safety Regulator?

Principal Accountable Persons or Accountable Persons must submit mandatory occurrence notices and reports to the Building Safety Regulator for the parts of the high-rise/higher-risk residential building they are responsible for and which are occupied as residential premises.

Principal Designers and Principal Contractors must submit mandatory occurrence notices and reports during

- construction of a new higher-risk building
- work on an existing higher-risk building, including work that causes it to stop being a higher-risk building
- work on an existing building that will make it a higher-risk building.

A higher-risk building is a building that has at least

- seven storeys or is at least 18 m high
- two residential units, or is a hospital or a care home.

When must a mandatory occurrence notice be submitted?

A mandatory occurrence notice must be submitted as soon as a safety occurrence has been identified, and before any report is submitted.

The Building Safety Regulator will request the following information in respect of the notice

- the high-rise residential building registration reference or the building control application reference
- contact details of the Accountable Person or person submitting the notice
- the date and time the safety occurrence was identified
- a brief description of the safety occurrence
- any immediate actions that have been taken since to keep people safe.

When a mandatory occurrence notice has been submitted, the Building Safety Regulator will issue a reference number that must be used to submit a mandatory occurrence report.

What is a mandatory occurrence report?

A mandatory occurrence report must be submitted after submitting the notice and within 10 calendar days of identifying the safety occurrence.

The Building Safety Regulator will request information regarding

- the mandatory occurrence notice reference
- who submitted the notice – if someone else submitted the notice, the Building Safety Regulator will need the contact details
- the type of safety occurrence that is being reported
- what happened or has the potential of happening
- what caused the safety occurrence and how it was discovered
- who is involved and the effect or potential effect on them
- what has been done and what is planned to keep people safe
- any other information that should be shared with others to add to everyone's learning
- any supporting information, such as documents, videos, photos.

The mandatory occurrence report can be submitted at the same time as the mandatory occurrence notice if all the relevant information is available.

During the design and construction phases of a higher-risk building, who is responsible for setting up a mandatory occurrence reporting procedure?

The Principal Designer is responsible for setting up a reporting system during the design stage of the project.

The notices and reports relate to the building that is under construction or undergoing building work. They do not relate to safety incidents relating to the construction site in general or any temporary structures. These types of incidents or accidents are the responsibility of the Principal Contractor and relevant contractor or sub-contractor and must be reported where applicable under the Reporting of Injuries, Diseases and Dangerous Occurrences Regulations 2013 (RIDDOR), as is usual practice.

There may be occasions when a RIDDOR notification will also need to be reported to the Building Safety Regulator as a mandatory occurrence report. For instance, the incident involves a structural collapse on site that if not remedied could cause death or serious injury to a significant number of people when the building is in use.

A mandatory occurrence reporting system must be established so that others can report safety occurrences to the Principal Designer or Principal Contractor for review.

Do the Principal Designer and Principal Contractor each operate their own mandatory occurrence reporting system?

The Principal Designer and Principal Contractor must

- establish and operate a mandatory occurrence reporting system before construction or building work begins
- maintain and operate their system throughout the project.

Principal Designers and Principal Contractors can operate the same or separate mandatory occurrence reporting systems.

Any mandatory occurrence reporting system must enable prompt reporting by

- designers and contractors working on the project
- others who periodically visit the work site.

The requirement for the mandatory occurrence reporting system and how it is envisaged it will work on the project should be discussed at design team meetings and pre-contract meetings. A coordinated approach, where everyone knows what needs to be done and by whom, will ensure that confusion does not reign on the site.

The Client should be instrumental in ensuring that a mandatory occurrence reporting system is established and must be satisfied that any incidents will be reported quickly and efficiently and that all duty holders will meet their legal duties.

Is the need for a mandatory occurrence reporting system applicable only to new build higher-risk buildings?

No. A mandatory occurrence reporting system must be established for all projects that involve work on an existing higher-risk building, unless the work consists only of exempt work or emergency repairs and work to an existing building that either makes it a higher-risk building, or stops it being a higher-risk building.

Refurbishment projects must be given as much attention as new build projects and clients are responsible for ensuring adequate controls are in place.

When is a mandatory occurrence report submitted to the Building Safety Regulator?

When a safety occurrence happens or is identified that meets the criteria of what to report to the Building Safety Regulator, the relevant person must submit

- a notice as soon as possible
- a report within 10 days of the safety occurrence being identified.

If all the relevant information is not available at the time of submitting the report, further details can be forwarded to the Building Safety Regulator as soon as is reasonably practicable.

What happens after the report is submitted to the Building Safety Regulator?

The Building Safety Regulator will review the report and determine the severity of the safety occurrence and will also review any safety measures that are in place and if they are sufficient to manage the risk.

The Building Safety Regulator will contact the person making the report if they

- need further information
- decide the incident or risk did not need to be reported
- need to start an investigation.

During an investigation, the Building Safety Regulator will review the findings and decide if any action is needed.

Does the Building Safety Regulator charge for dealing with mandatory occurrence notices and reports? And if so, who pays?

There are no charges for making a mandatory occurrence notice and report, but investigations do incur cost recovery charges.

Costs as of 2025 are: £151/h worked by Building Safety Regulator staff; plus any costs to the Building Safety Regulator or relevant authorities and third parties – for example, fire safety services, structural engineers.

The person who pays any charges will be

- the Client

or

- the Principal Accountable Person if a certificate has been issued for the building/relevant part of the building in accordance with Regulation 6(2).

A dangerous occurrence has happened on site and the Principal Contractor has submitted a notification under RIDDOR. Will a mandatory occurrence notice and report have to be submitted to the Building Safety Regulator?

If the incident on site meets the criteria for a mandatory occurrence notice than yes, it will.

As an example:

During building work, part of a building collapses and injures people. The Principal Contractor reports the incident under RIDDOR. The Principal Contractor also submits a mandatory occurrence notice and report, as the incident involves structural failure, which if not remedied risks death or serious injury to a significant number of people when the building is in use.

It is important to remember that there are two reporting systems that may need to be operated on a construction site that is involved in constructing higher-risk buildings and clear procedures must be in place to ensure that the legal duties on all parties are fulfilled.

The Project Plan and Construction Phase Plan should include clear procedures as to who is responsible for carrying out the activities needed to satisfy the law.

REFERENCES

Building Act 1984.
Building Safety Act 2022.
The Building (Higher-Risk Buildings Procedures) (England) Regulations 2023.
The Building (Public Bodies and Higher-Risk Building Work) (England) Regulations 2023.
The Building Safety (Registration of Higher-Risk Buildings and Review of Decisions) (England) Regulations 2023.
The Higher-Risk Buildings (Descriptions and Supplementary Provisions) Regulations 2023.
The Higher-Risk Buildings (Keeping and Provision of Information etc.) (England) Regulations 2024.
The Higher-Risk Buildings (Key Building Information etc.) (England) Regulations 2023.
The Higher-Risk Buildings (Management of Safety Risks etc.) (England) Regulations 2023.
Various guidance documents on Gov.uk website: www.gov.uk.
Various guidance documents on HSE/Building Safety Regulator website: www.hse.gov.uk.

Mandatory occurrence notice (Example)

<table>
<tr><td colspan="4">Submit a mandatory occurrence notice and report when a safety occurrence has caused, or is likely to cause:

• the death of a significant number of people
• serious injury of a significant number of people

A safety occurrence is an incident involving, or risk that could cause:

• structural failure of the building
• the spread of fire or smoke in the building

A safety occurrence is something which if not remedied, could cause serious harm to people when the building is in use</td></tr>
<tr><td>1</td><td colspan="3">Your name (enter your answer)</td></tr>
<tr><td>2</td><td colspan="3">Your address (enter your answer)</td></tr>
<tr><td>3</td><td colspan="3">Contact telephone number (enter your answer)</td></tr>
<tr><td>4</td><td colspan="3">Email address (enter your answer)</td></tr>
<tr><td>5</td><td>Position (please tick)</td><td>☐ Principal Designer</td><td>☐ Principal Contractor</td></tr>
<tr><td>6</td><td colspan="3">Building address at which the safety occurrence took place (enter your answer)</td></tr>
<tr><td>7</td><td colspan="3">Building control application reference number (enter your answer)</td></tr>
<tr><td>8</td><td>Does the safety occurrence relate to spread of fire or structural safety (please tick)</td><td>☐ Spread of fire</td><td>☐ Structural safety</td></tr>
</table>

9	Date and time of occurrence (enter your answer)
10	Details of the safety occurrence (enter your answer)
Date Notice submitted (enter your answer)	

Mandatory occurrence report (Example)

This Report must be submitted within **TEN DAYS** of the date of the Mandatory Occurrence Notice	
Mandatory Occurrence Notice reference number	
Building Registration/Building Control reference number	
Building address where safety occurrence incident happened	
Building registration number if applicable	
Contact details	Name
	Email
	Contact telephone number
Name of Organisation	
Position or involvement in the Project	
The incident or risk that has happened i.e. describe the safety occurrence (tick which applies)	☐ Structural ☐ Fire ☐ Both
When did it occur?	

How was it discovered?	
Details of what happened to cause the safety occurrence	
Who is involved in the safety occurrence?	
What effect will the safety occurrence have, or what potential risk is posed?	
What has been done to keep people safe?	
What else needs to be done to keep people safe?	
Please add any supporting information – please tick all those that apply	☐ Documents ☐ Videos ☐ Photos ☐ Investigation reports ☐ Other
Signature of person submitting the Report	
Date of submission	

The Building Safety Act 2022: Questions and Answers: A practical approach to design and management of higher-risk buildings

Pat Perry
ISBN 978-1-83662-467-7
https://doi.org/10.1108/978-1-83662-464-620251019

Chapter 17
The new change control regime

What is the change control regime introduced by the Building Safety Act 2022 for higher-risk buildings?

A new change control regime has been introduced under the Building Safety Act 2022 in relation to higher-risk buildings applicable during the design and construction phase of a project. It has been implemented through the Building (Higher-Risk Buildings Procedures) (England) Regulations 2023 (HRB Procedures Regulations), which came into force on 1 October 2023.

The new change control process will ensure that all major changes to designs, construction processes and construction materials invariably made during the construction phase have been checked and verified by a competent authority.

Why has a new process been implemented?

Dame Judith Hackitt was commissioned by the government in 2018 to carry out an independent review of building regulations and fire safety following the Grenfell Tower fire in 2017.

She concluded that the 'current system of building regulations and fire safety is not fit for purpose and that a culture change is required to support the delivery of buildings that are safe, both now and in the future'.

Her recommendation in respect of changes that occur to higher-risk buildings was: 'A stronger change control process that will require robust record-keeping by the duty holder of all changes made to the detailed plans previously signed off by the JCA. More significant changes will require permission from the JCA to proceed.'

Building Control departments often have no records of any design changes made during the progress of the construction works on site, nor any records of changes in materials used. This means that building standards could be less than those approved and these could lead to major weaknesses in building and fire safety expectations.

What is required in respect of the new change control process?

Part 3 of Building (Higher-Risk Buildings Procedures) (England) Regulations 2023 requires the Client to record and revise agreed documents to reflect the changes made to a design already submitted to the Building Safety Regulator. It also requires an application to the Building Safety Regulator to approve certain controlled changes that are proposed in respect of higher-risk buildings after building control approval has already been given. Such changes include deviations from the approved plans, changes to a stage of higher-risk building work and certain changes to agreed documents.

Regulation 26 sets out and defines three change categories: major, notifiable and recordable.

What is a 'controlled change'?

Regulation 18 Building (Higher-Risk Buildings Procedures) (England) Regulations 2023 states that a controlled change is when one or more of the following is proposed

- a change to any current plans of any work or proposed work or the carrying out of work otherwise than in accordance with the current plans
- a change to any stage of higher-risk building work, including adding or removing a stage
- a change to any strategies, policies or procedures described in any current agreed document.

What is a major change?

A 'major' change is a change that would undermine the basis by which the building control approval was granted. This could be something such as a change to the height of a proposed higher-risk building and would require a Client to obtain prior Building Safety Regulator approval. The process involves submission of a change control application in prescribed form, signed by the Client and containing

- a description of the change
- reasons for making it
- details of the advice sought in relation to that change
- an assessment of the agreed documents affected by the proposed change
- an explanation of how the building will be compliant once the change has been made.

What examples are there of major changes that need to be notified to the Building Safety Regulator?

Regulation 26 Building (Higher-Risk Buildings Procedures) (England) Regulations 2023 sets out some examples as follows

- a material change of use to any part of a proposed or existing higher-risk building
- a change that increases or decreases the external height or width of a proposed or existing higher-risk building
- a change to the number of storeys in a proposed or existing higher-risk building
- a change to the structural design or structural loading of the building
- a change to the number or width of the staircases in a proposed or existing higher-risk building or to the length, number or width of any escape route
- a change to the external wall of a proposed or existing higher-risk building including a wall tie, wall restraint fixing or support system in the wall

- a change to active or passive fire safety measures in a proposed or existing higher-risk building referred to in the Fire and Emergency File
- a change to the layout or dimensions of the common parts within a proposed or existing higher-risk building.

A major change has been notified to the Building Safety Regulator. Can we start or continue with the construction works?

No.

Approval from the Building Safety Regulator must be obtained before the work relating to that major change begins.

The Building Safety Regulator must determine the application within 6 weeks from receipt or within such longer period as the Building Safety Regulator and applicant agree in writing.

Construction works could be significantly held up if multiple design changes are made and this may have implications for contract negotiations, accountability, liability for delays, and so forth.

What is a notifiable change?

A Client must notify the Building Safety Regulator of a notifiable change in accordance with Regulation 20(1) Building (Higher-Risk Buildings Procedures) (England) Regulations 2023 before the work relating to that change starts, but does not need the Building Safety Regulator's approval before commencing construction.

A notifiable change is defined in Regulation 26 of the above Regulations may include

- a change to the Construction Control Plan, Change Control Plan, partial completion strategy or staged work statement
- a change to the wall tie, wall restraint fixing or support system in any (or any proposed) wall (excluding an external wall)
- a change of any construction product or building element that has been (or is proposed to be) used in or on a proposed higher-risk building, such that the replacement is of an equivalent or higher classification under the reaction to fire classification
- any other change to the Fire and Emergency File or the fire safety compliance information.

A notifiable change has been made to the Building Safety Regulator. Can we start or continue with the construction works?

Yes.

Once the Building Safety Regulator has been informed of the notifiable change, works can continue on site.

What is a recordable change?

A recordable change is a minor change that does not need to be raised with the Building Safety Regulator.

Details of a recordable change should, however, be recorded in the change control log.

Who is responsible for managing the change control process?

In the event that a change needs to be made from the original scheme, then it will be the responsibility of the Client, Principal Designer and Principal Contractor to decide whether the changes are recordable, notifiable or major and to record the details of the change within the change control log.

Can the Building Safety Regulator change the category of notification once all the documents have been received?

Yes.

If the Building Safety Regulator considers that the wrong decision has been made regarding the category of change notified, they have the power to change the category. This will mean that if the change is from notifiable to major, all works relating to the change proposed will have to stop until such time as the Regulator has approved the proposed changes.

Who is responsible for creating the change control log?

The Principal or Sole Contractor is responsible for creating the change control log and for ensuring that it is kept up to date.

A practical way of ensuring that the log is kept up to date is to include an item 'Change Control Log' on every progress meeting/project meeting agenda so that it can be reviewed by the full project team and any additions or omissions can be highlighted. The design team might take responsibility for asking questions at the meeting, such as 'Are there any design or construction changes we need to be aware of and that may need to be notified to the Building Safety Regulator?'

The Client/project manager may do likewise.

What is the change control log?

The details of changes must be recorded within the change control log for the purposes of the documentation requirements of the new building safety regulations.

The Building (Higher-Risk Buildings Procedures) (England) Regulations 2023 Regulation 19(2) set out the following information that must be recorded as

- the name of the Client, Principal Designer or Principal Contractor recording the change
- a description of the proposed change

- details explaining the reasons why the change has been proposed
- identification of whether the change is recordable, notifiable or major
- if applicable, the details of any persons whose advice was sought in relation to the change, including a summary of their advice
- an assessment of which agreed document has been affected by the change and confirmation that a revised version has been produced
- an explanation of how the change will, once carried out, comply with the applicable building regulations (a 'compliance explanation').

Overall, the change control process will ensure that any substitution that takes place is done in line with the performance requirements of the specification.

Is there a specific process to follow when notifying the Building Safety Regulator of a notifiable change to the proposed design or construction works?

Yes.

Regulation 18(3) states that where a change is deemed to be a notifiable change, the Client must notify the Regulator in accordance with Regulation 20. This, in turn, states that the application must include

- a description of the proposed controlled change
- an explanation of the reasons why the change has been proposed
- a list of any individuals whose advice was sought in relation to the proposed change and a summary of that advice (such advice should, of course, be from those competent to give that advice and ideally be in the form of a report)
- a compliance explanation in relation to the proposed change
- an assessment of the agreed documents affected by the proposed change
- a statement signed by the Client confirming agreement to the application being made and that the information contained in the application is correct
- the revised version of the agreed document produced in accordance with Regulation 18(2)(b).

Is there a specific process to follow when notifying the Building Safety Regulator of a major change to the proposed design or construction works?

Yes.

Regulation 21 of the Building (Higher-Risk Buildings Procedures) (England) Regulations 2023 sets out what must be submitted when a 'major change' change control application is made.

The application must be in writing and be signed by the Client as applicant.

It should include (among other things)

- the name, address, telephone number and (if available) email address of the Client
- the name, address, telephone number and (if available) email address of the Principal Contractor (or Sole Contractor) and the Principal Designer (or Sole or Lead Designer)
- a statement that the application is made under this regulation
- a description of the proposed controlled change.

The application must be accompanied by

- an explanation of the reasons why the change has been proposed
- a list of the name and occupation of each person, if any, whose advice was sought in relation to the proposed change and a summary of any advice provided
- where the change is to the number of flats or residential rooms, plans showing the number of flats or residential rooms it is proposed the higher-risk building will contain
- an assessment of the agreed documents affected by the proposed change
- a compliance explanation in relation to the proposed change
- a statement, where the application is made by someone on behalf of the Client, signed by the Client confirming they agree to the application being made and that the information contained in the application is correct
- the application must be accompanied by the revised version of the agreed document produced.

If a change control application does not comply with the requirements set out in Regulation 21, it will not be valid. Where this is the case, the Building Safety Regulator will notify the applicant and provide reasons. The project will be delayed as no works relating to any proposed changes can be carried out until approved by the Regulator.

What are the options for the Building Safety Regulator when considering a change control request?

The application must be determined within 6 weeks of receipt.

The Regulator has the option to consult other statutory bodies – for example, fire authority, local authority, water/sewerage authority, specialist structural engineers and others as deemed necessary.

All consultees must be given appropriate information and sufficient time to consider the details of the change requested.

The ultimate aim of the change control process is to ensure that the building, together with any changes from the originally submitted information, will meet all the requirements of the Building Regulations and the on-going occupation requirements of the Building Safety Act 2022.

The Regulator can determine that a notifiable request for change is not a notifiable request but a major change and can reject the application.

Applications will be rejected if they do not meet the requirements of Regulation 21 of the Building (Higher-Risk Buildings Procedures) (England) Regulations 2023.

Lack of sufficient information will lead the Regulator to request further details from the Client and this may add to project delays as the Regulator may need more than 6 weeks to consider the application once all of the information has been received.

The Regulator will reject the application if the details provided do not demonstrate that the proposed changes meet the requirements of the Building Regulations.

Reasons for rejection must be given to the Client in writing. The applicant will have the option to review/appeal the decision.

Does the Building Safety Regulator levy costs for dealing with the change control process?

Yes.

As with nearly all activities carried out by the Building Safety Regulator, there will be costs payable by the Client.

Costs are laid out in the document 'Building Safety Regulator charging scheme', published by The Building Safety Regulator/Health and Safety Executive.

Submission of a change control application will be a £189 (as of April 2025) charge with the application, plus £151/h (as of April 2025) worked by Building Safety Regulator staff, plus any costs to the Building Safety Regulator or relevant authorities and third parties such as during any consultation.

Are there any criminal offences associated with any failures to follow the change control process?

Yes.

The Building Safety Regulator can issue either compliance notices or stop notices whenever it becomes aware that the new building safety laws are being abused. These notices are especially issued where evidence has been obtained that the Building Regulations are being breached or that specific sections of the Building Safety Act 2022 and any of its secondary legislation are being contravened.

Should either the compliance notice or the stop notice not be complied with, the recipient could be prosecuted – that is, criminal offences laid before the appropriate court.

Fines on conviction will be unlimited or 1- or 2-year prison sentences could be levied, depending on the level of court hearing the case – for example, Magistrates' Court or Crown Court.

TOP TIPS

- Plan in advance to ensure compliance with the Building Safety Act 2022.
- Formalise the project team and ensure everyone knows their responsibilities.
- Nominate one person to be the lead on change control processes.
- Prepare the necessary paperwork before the project starts
  - Change Control Plan
  - Change Control Log.
- Check everyone meets competency standards.
- Coordinate designs to reduce likelihood of post commencement changes to the design or construction process or materials.
- Understand what changes are
  - recordable
  - notifiable
  - major.
- Keep records and documentation up to date.
- Coordinate design discussions – involve all disciplines.
- Remember – you must always confirm compliance with Building Regulations.
- Failure to comply with legal duties is a criminal offence – unlimited fine or up to 2 years imprisonment.

REFERENCES

Building Regulations 2020 (as amended).

Building Safety Act 2022.

The Building (Higher-Risk Buildings Procedures) (England) Regulations 2023.

Guidance documents

Building Control: An overview of the new regime: Gateways 2 and 3 – HSE August 2023.

Change Control Log (Example)

<table>
<tr><td>Name of project:</td><td>Change control reference:</td><td>Date:</td></tr>
<tr><td colspan="3">Address of project:</td></tr>
<tr><td>Name of person recording the current change proposed:</td><td colspan="2">Position of person recording the change:</td></tr>
<tr><td colspan="3">What is the proposed change? (please describe in as much detail as possible)</td></tr>
<tr><td colspan="3">Why has the change been proposed?</td></tr>
</table>

<table>
<tr><td>Is the proposed change (please tick)</td><td>☐ Recordable
☐ Notifiable
☐ Major</td></tr>
<tr><td rowspan="3">Who has been consulted regarding this change?</td><td>Name:</td></tr>
<tr><td>Discipline:</td></tr>
<tr><td>Occupation:</td></tr>
<tr><td colspan="2">What advice/information did they give?</td></tr>
<tr><td colspan="2">Which documents will need to be updated to include for the changes proposed?</td></tr>
<tr><td colspan="2">How and when will the documents be updated and by whom?</td></tr>
</table>

Describe how the proposed changes to the design, construction processes or construction materials will meet the requirements of the Building Regulations and the Building Safety Act 2022 for higher-risk buildings:		
Any other relevant information		
Building Safety Regulator Ref. No.		
Building Regulations/Control Application No.		
Client sign off	Name:	Date:
Principal Designer sign off	Name:	Date:

Change Control Plan (Example)

Suggested contents:

Introduction and Approach:

Include details about the project

Include details about the overall change control management approach

Explain why changes might have to occur

Describe any software in use to record and manage changes

Describe how changes will be raised, processed and agreed

Roles and Responsibilities:

Include details of the Client

Include details of all of the project team, disciplines etc

Include details of each person's duties

Describe who can propose changes and when

Project Specifications:

Describe the rationale behind the original design and the standards/regulations that are met

Describe details/specifications that cannot be changed

Describe the records to be kept recording changes

Change Control Assessment:

Describe how changes will be considered and by whom

Describe how the information will be communicated to other parties

Describe how other experts will be consulted

Describe how changes will be signed off

Describe how changes will be documented

Describe how the proposed changes will be tracked

Change Request Confirmation:

Describe who will ultimately sign off the change

Ensure signatures and dates

Ensure how statement to comply with the regulations is obtained

Describe how all change request records will be kept and where

The Building Safety Act 2022: Questions and Answers: A practical approach to design and management of higher-risk buildings

emerald PUBLISHING ice Publishing

Pat Perry
ISBN 978-1-83662-467-7
https://doi.org/10.1108/978-1-83662-464-620251020

Chapter 18
Residents' engagement strategy

What is a residents' engagement strategy?

The residents' engagement strategy sets out how residents and owners of residential units are

- involved in building safety decisions
- informed about building safety decisions.

Building safety decisions are made by an Accountable Person and are about

- the management of the building
- management of building safety risks
- any decision connected to the duties of an Accountable Person.

A resident is anyone over 16 years of age who lives in or owns a residential unit.

Why is a residents' engagement strategy necessary?

The Independent Review of Building Regulations and Fire Safety, conducted by Dame Judith Hackitt, found that residents did not have a strong enough voice in the management of their buildings.

The residents of Grenfell Tower had raised many concerns about the safety of their tower block but both the landlord and the managing agents failed to acknowledge those concerns and no action on critical defects was taken.

The Building Safety Act 2022 has introduced clearer routes for resolving issues raised by residents about their safety. Residents now have a chance to participate in decision making as well as have access to key building safety information.

Under the more stringent higher-risk regulatory regime, the Building Safety Act sets out a new duty holder role for the safety management of higher-risk residential buildings, known as the Accountable Person.

Is a residents' engagement strategy legally required under the Building Safety Act 2022?

Yes.

Section 91 of the Act sets out the requirements for a residents' engagement strategy as follows

(1) The principal accountable person for an occupied higher-risk building must—

(a) as soon as reasonably practicable after the relevant time prepare a strategy (a "residents' engagement strategy") for promoting the participation of relevant persons in the making of building safety decisions;
(b) review the strategy at prescribed times, and revise it if they consider it necessary or appropriate to do so;
(c) in prescribed circumstances, consult relevant persons and prescribed persons on the strategy and take any representations made on the consultation into account when next reviewing the strategy;
(d) act in accordance with the strategy.

(2) "Building safety decision" means a decision by an accountable person for the building that—

(a) is about the management of the building, and
(b) is made in connection with the performance of a duty of the accountable person under, or under regulations made under, this Part.

(3) The strategy must include information about—

(a) the information that will be provided to relevant persons about decisions relating to the management of the building,
(b) the aspects of those decisions that relevant persons will be consulted about,
(c) the arrangements for obtaining and taking account of the views of relevant persons, and
(d) how the appropriateness of methods for promoting participation will be measured and kept under review.

(4) As soon as reasonably practicable after the strategy is prepared or revised, each accountable person for the building must give a copy of the strategy to—

(a) each resident of the building who—

(i) is aged 16 or over, and
(ii) resides in a residential unit in the part of the building for which the accountable person is responsible;

(b) each owner of a residential unit in that part of the building;
(c) any prescribed person.

(5) The duty under subsection (4)(a) does not apply in relation to a resident if the accountable person—

(a) is not aware of the resident, and
(b) has taken all reasonable steps to make themselves aware of persons who reside in residential units in the part of the building for which the accountable person is responsible.

(6) The Secretary of State may by regulations—

(a) make further provision about the content of a residents' engagement strategy;
(b) make provision about the preparation, review or revision of a residents' engagement strategy, in cases where there is more than one accountable person for the building;
(c) make provision about the way in which a copy of the strategy is to be given;
(d) make provision about consultations under this section.

(7) In this section—

"relevant persons" are—

(a) residents of the higher-risk building who are aged 16 or over, and
(b) owners of residential units in the building;

"the relevant time" has the meaning given by section 85 BSA 2022.

Who has to draw up the residents' engagement strategy?

The Principal Accountable Person is responsible and must

- prepare a residents' engagement strategy
- act in accordance with the strategy
- review and revise the strategy and keep a record of the reviews
- provide the latest version to each Accountable Person
- when necessary, consult residents and Accountable Persons about the strategy and take their opinions into account
- tell residents about building safety work resulting from a building safety decision.

Accountable Persons must

- provide the latest version of the strategy to the residents in their parts of the building
- work with any other Accountable Persons for the building to create and update the strategy.

When is a residents' engagement strategy required?

The Principal Accountable Person should prepare a strategy as soon as possible when

- the building is or becomes occupied
- they become the Principal Accountable Person.

Is a residents' engagement strategy required for higher-risk buildings that are already in existence or just for new build premises?

A residents' engagement strategy is required for all higher-risk buildings, including those already occupied by residents.

New build higher-risk buildings require a residents' engagement strategy as soon as the building becomes occupied.

Does the residents' engagement strategy need to be submitted to the Building Safety Regulator?

Yes, but timescales for submission can vary.

All existing higher-risk residential buildings had to be registered with the Building Safety Regulator by 1 October 2023. Each application has to be accompanied by certain information about the building.

The Regulator reviews all the applications and, provided sufficient information has been received, will register the building.

Once a building has been registered, the Building Safety Regulator will prioritise those deemed to be the highest risk and will inform the Principal Accountable Person that they must apply for a Building Assessment Certificate.

Within 28 days of the Building Safety Regulator request, an application must be made for a Building Assessment Certificate. Specific information has to accompany the application and a residents' engagement strategy forms part of the pack of information required.

What needs to be included in the residents' engagement strategy?

Residents must be involved in decisions about their building and must fully understand the process of how decisions are made and what influence they can have.

A residents' engagement strategy should include

- how residents will be involved in decisions about their building's safety, particularly during any refurbishment
- the strategic approach to communicating with residents, including the channels and forums through which different information will be shared
- what steps will be taken to ensure the engagement takes account of the diverse needs of residents
- how core information about building safety will be shared proactively with residents
- how appropriate detailed information about building safety will be made available to residents on request
- how complaints about safety will be handled effectively and efficiently
- how residents will be informed about their own safety responsibilities
- how these responsibilities will be managed
- how implementation of the strategy will be measured.

Are there any other legal duties in respect of the residents' engagement strategy?

A key requirement of the Building Safety Act 2022 is for the Principal Accountable Person/Accountable Person to develop a formal complaints procedure that can be clearly understood by all residents and effectively used. Residents must be able to raise a formal complaint about a 'relevant' issue concerning the building they live in. This is to be known as a relevant complaint.

A relevant complaint involves

- structural failure and spread of fire (building safety risks)
- the performance of an Accountable Person.

Relevant complaints must be investigated by the Principal Accountable Person/Accountable Person.

Residents can make a complaint directly to the Building Safety Regulator if they believe that they

- cannot make a complaint
- feel unable to raise their concerns
- are not satisfied with the response from the Principal Accountable Person
- are not satisfied with the final response
- have unresolved issues
- have not got a response.

The Building Safety Regulator will review the complaint and if it is deemed a relevant complaint they will investigate and take matters up with the Principal Accountable Person.

What happens after the residents' engagement strategy has been drafted?

The Principal Accountable Person or Accountable Person must give a copy to

- each resident of the building who
  - is aged 16 or over, and
  - resides in a residential unit in the part of the building for which the accountable person is responsible
- each owner of a residential unit in that part of the building
- any prescribed person.

Does the residents' engagement strategy need to be reviewed and updated regularly?

Yes.

The Principal Accountable Person must keep the strategy document up to date and review it

- at least every 2 years
- after submitting a mandatory occurrence report to the Building Safety Regulator
- after the completion of significant material alterations to the building.

The Principal Accountable Person must record each review, whether or not they make changes to the strategy.

How can the strategy be communicated to all residents?

The consultation process used to draw up the strategy should include discussions as to how residents would prefer to receive the information they are legally required to have.

There are many different communication models available but the law requires that every resident has an individual copy.

This could be by way of a printed document hand delivered to every residential unit, an electronic copy attached to an email or a digital version on the community website.

Versions will need to be available in the various languages spoken by the residents and consideration must also be given to formats suitable for visually impaired residents, those with learning difficulties or other mental health conditions.

Include discussions on building safety in any community group activities running within the building, whenever social get togethers happen, food bank facilities or whenever and wherever residents meet together.

Some housing providers have formed a residents' engagement panel to meet regularly to discuss matters of concern within the building(s) and to agree actions to be taken, and so on.

Remember: Principal Accountable Persons/Accountable Persons are legally responsible for acting on major concerns in respect of building safety matters especially in respect of fire safety and structural safety.

REFERENCE

Building Safety Act 2022 – sections 91 to 94.

The Building Safety Act 2022: Questions and Answers: A practical approach to design and management of higher-risk buildings

emerald PUBLISHING ice Publishing

Pat Perry
ISBN 978-1-83662-467-7
https://doi.org/10.1108/978-1-83662-464-620251021

Chapter 19
Construction products

Why are construction products important in relation to the Building Safety Act 2022?

In his report at the end of Phase I of the Grenfell Tower Inquiry, Sir Martin Moore-Bick concluded that the principal reason why fire spread so rapidly vertically up and around the tower was the presence of aluminium cladding panels with polyethylene cores. These acted as a source of fuel, with molten, burning polyethylene dripping down the face of the building igniting fires lower down. A further contribution was made by polyisocyanurate (PIR) and phenolic foam insulation boards behind the panels, and possibly by the components and detailing of the window surrounds.

The horizontal spread of fire was the result of fire breaking back into the building through the inevitable failure of the glass in the windows, with compartmentation within the building then being lost as a consequence of kitchen extractor fans deforming and some fire doors failing to close.

He also concluded that 'there was compelling evidence that the external walls of the building failed to comply with' the Building Regulations.

So, irrespective of any issues of accountability, the conclusion reached was that products that were inappropriately selected, or that were wrongly installed, or that failed to perform as they should, and/or that should not have been approved for use in the first place were implicated in the tragedy.

Subsequently, in Phase 2 of the Inquiry, evidence was presented suggesting default on the part of those manufacturing and assessing construction products, severely undermining confidence in the system by which such products are regulated and come to market. These allegations fall broadly into four categories.

(a) Failure by manufacturers to disclose all information relevant to the product and its assessment.
(b) Failure on the part of those responsible for testing to follow proper procedures in conducting the assessment process.
(c) Failure to ensure that certificates or classification reports were properly supported by the assessment process.
(d) Failure by manufacturers to ensure that performance claims made for their products were limited to those that were supported by the assessment process and were not misleading.

In Dame Judith Hackitt's final report 'Building a Safer Future', published in May 2018, she stated that

> The system that covers product testing, labelling and marketing is at least as complicated as the entire regulatory system which was mapped in the interim report... it is apparent that the current system makes it difficult to know whether the right products are being used.

The report then went on to make some recommendations that would set a direction of travel for improved product safety.

She has also stated: 'It should be self-evident to everyone that we need products which do the job they are expected to do and are marketed honestly and transparently'.

From the in-depth inquiries commissioned after the Grenfell Tower fire, it can be concluded that construction products play a significant role in the safety of buildings and that inferior or poorly manufactured products and those used in locations and for purposes that were not anticipated can cause significant safety failings.

The government agreed with the Public Inquiry findings and also with the recommendations from the Hackitt Review and commissioned an independent review to address the following question.

> How should the UK system for testing the safety of construction products and the use of data from the system be strengthened, to inspire confidence that those products are safe and perform as labelled and marketed when incorporated into construction work?

The report is known as the Independent Review of the Construction Product Testing Regime.

What is a construction product?

A 'construction product' is not defined explicitly in the Building Safety Act 2022 but section 148(9) links the definition to existing domestic and EU regulations. The main definition is found in the Construction Products Regulations 1991: 'any product which is produced for incorporation in a permanent manner in works'.

This is a broad definition and is likely to encompass most materials used in construction.

How will the Building Safety Act 2022 affect construction products?

Section 146 and Schedule 11 of the Building Safety Act give the Secretary of State wide-ranging powers to make construction product regulations that are intended to include, among other things

- the introduction of a general safety requirement prohibiting construction products from being placed on the market unless they are safe products, meaning that they do not present a risk to the health and safety of people, or the risk is as low as possible
- an obligation for manufacturers to carry out a risk assessment in order to comply with the general safety requirement described above

- the creation of a new category of 'safety-critical' construction products, defined in Schedule 11 as those where 'any failure of the product would risk causing death or serious injury to any person', and more stringent regulation of such products
- the creation of new criminal offences and the provision for enforcement actions and penalties for non-compliance with such regulations.

What has the Secretary of State done in respect of construction product regulations?

The Office for Product Safety and Standards (OPSS) was appointed the UK Construction Products Regulator for construction products in 2021. Its purpose is to ensure the claimed performance requirements of construction products are met, thus enabling safer, better buildings across the UK.

The OPSS regulates through enforcement of

- Construction Products Regulations 2013
- Construction Products (Amendment) Regulations 2022.

The UK government adopted new Construction Product Regulations issued by the EU in 2024, which came in to force in January 2025.

The updated Regulations introduce key measures that manufacturers must follow to enhance compliance and sustainability.

- A major addition is the Digital Product Passport (DPP), which serves as a centralised digital record containing technical specifications and environmental data, improving traceability across the EU market.
- CE marking has also been expanded to include not only technical performance but also environmental impact, ensuring products adhere to both safety and sustainability standards.
- Manufacturers are now required to report on climate-related factors such as CO_2 emissions and energy consumption for key construction products, with additional indicators to be incorporated over time.

What are the requirements of section 148 of the Building Safety Act 2022?

Section 148 creates a new course of action that allows for manufacturers and sellers of construction products to be held accountable where the use of a construction product in works causes or contributes to a dwelling or a building containing a dwelling(s) being unfit for habitation on completion of those works.

Previously it was usual for no legal relationship to exist between the current owner or resident of a building and manufacturers of construction products. This left current owners/residents with no remedy for defective materials that rendered the building unfit for habitation. This was highlighted as a key issue in Dame Judith Hackitt's Independent Review, which stated that the 'principle of risk being owned and managed by those who create it … should extend to the safety of those who live in and use the products of the construction industry'.

For a manufacturer to be liable the following conditions must be met.

Condition A: A person either

- fails to comply with a construction product requirement (both prospectively and current) (a construction product in relation to a construction product requirement means a requirement
  - under Construction Products Regulations has the meaning specified in those regulations
  - under the EU-CPR 2011 has the meaning specified in that regulation
  - under the Construction Products (Amendment etc.) (EU Exit) Regulations 2019 has the meaning specified in that regulation)
- makes a misleading statement in relation to a construction product in its marketing or supply (this raises the question of whether a supplier would be held responsible for relying on and repeating a misleading statement made by a manufacturer)
- manufactures a construction product that is inherently defective.

Condition B: The construction product is installed in or applied to or attached to a relevant building during its construction or 'otherwise in relation to' the building. (The use of the term 'or otherwise in relation to' raises the question as to whether the government wishes to encompass refurbishment and renovation work.)

Condition C: When the works are completed, the relevant building or a dwelling contained within the building is unfit for habitation.

Condition D: The failure identified in Condition A was the cause or one of the causes of the building or dwelling being unfit for habitation.

A claim for compensation can be brought through the civil courts against a manufacturer or a supplier of construction products. This claim can be brought by those who have a relevant interest in the relevant building that contains a dwelling and have suffered a loss due to the building or dwelling(s) being unfit for habitation. (Loss can be claimed for personal injury, damage to property or economic loss, which includes the costs of remediation.)

The limitation period for claims brought under this section is 15 years (an extension to the standard 6 years) from when the cause of action accrues. This is in relation to buildings completed after 28 June 2022.

What are the details in Schedule 11 of the Building Safety Act 2022?

Schedule 11 of the Act outlines provisions relating to

- declarations of performance and how they are marketed
- statements and claims made about a product's performance
- product packaging and markings
- the information required for presentation (including risk information)

- the monitoring, assessment and verification of performance (including sample testing)
- the storage and transportation of products
- the taking of corrective action, including market withdrawal, product recall and information on whom the product has been supplied
- the recording and follow-up of complaints
- the production and retention of documentation and samples
- the notification of risk to relevant authorities and subsequent cooperation
- the appointment of authorised representatives.

Comparable to the consumer product requirements in the General Product Safety Regulations 2005, the Act will prohibit the marketing or supply of construction products deemed unsafe. Construction products will only be considered safe if, under ordinary conditions, the product does not present any reasonable risk to the health and safety of a person.

Which products are considered 'safety critical'?

Schedule 11 defines 'safety-critical products' as 'construction products which are included in a list contained in construction products regulations'.

Although the list has yet to be drawn up, the Act provides insight into what products may fall into the safety-critical category. The government will construct the list with the help of industry professionals and persons in reputable positions.

The Act, in its current state, indicates that construction products will be included in the list if

- any product failure would present a risk of death or serious injury (in the view of the Secretary of State)
- the product does not have designated standards for the 2011 Regulations.

Does the Building Safety Act 2022 change liability in respect of construction products?

Yes. There is a new liability applying to manufacturers and distributors of construction products to be held accountable where the product used in works causes or contributes to a dwelling being unfit for habitation. This applies retrospectively for cladding products only and where the relevant products are in breach of regulations, inherently defective or have been mis-sold. This could cause many issues as the manufacturer does not have control over where a product is being used.

New buildings and new construction works materials will be covered.

Claimants must show that the work carried out on the building was not undertaken in 'a workmanlike or professional manner' or that 'proper materials' were not used. This emphasises the importance for manufacturers and sellers of construction products to provide accurate, clear and unambiguous information about their products.

The Construction Products Association advises manufacturers to examine the benefits of engaging with the Code for Construction Product Information (CCPI).

What liability do manufacturers and suppliers have in respect of cladding materials?

Section 149 of the Building Safety Act 2022 makes provision for new and retrospective statutory liability where a cladding product used in works causes, or is a factor in causing, a building or a dwelling it contains being unfit for habitation following completion of the works.

As with section 148, property owners/residents previously had no remedy for defective materials that rendered the building or its contained dwelling(s) unfit for habitation.

In 'Building a Safer Future', Dame Judith Hackitt stated that the 'principle of risk being owned and managed by those who create it ... should extend to the safety of those who live in and use the products of the construction industry'.

For the manufacturer to be liable, certain criteria must be met, such as

- fails to comply with a cladding product requirement (retroactively, prospectively and currently)
- makes a misleading statement in relation to a cladding product in its marketing or supply
- manufactures a cladding product that is inherently defective
- the cladding product is attached to, or included in, the external wall of a relevant building during the course of works to that building or 'otherwise in relation to' the building
- when the works to install the cladding product are completed, the relevant building or a dwelling within the building is unfit for habitation.

In relation to buildings completed after 28 June 2022, the limitation period for claims brought under this section is 15 years (an extension to the standard 6 years) from when the cause of action accrues.

In relation to buildings completed **prior** to 28 June 2022, the limitation period for claims brought under this section is 30 years from when the cause of action accrues. Claims can therefore be brought in respect of works completed from mid-1992 onwards.

Pat Perry
ISBN 978-1-83662-467-7
https://doi.org/10.1108/978-1-83662-464-620251023

Index

Printed and bound by CPI Group (UK) Ltd, Croydon, CR0 4YY

24/06/2026

14906982-0005